PHILOSOPHY OF PHYSICS

SYNTHESE LIBRARY

MONOGRAPHS ON EPISTEMOLOGY,

LOGIC, METHODOLOGY, PHILOSOPHY OF SCIENCE,

SOCIOLOGY OF SCIENCE AND OF KNOWLEDGE,

AND ON THE MATHEMATICAL METHODS OF

SOCIAL AND BEHAVIORAL SCIENCES

Editors:

DONALD DAVIDSON, *The Rockefeller University and Princeton University*

JAAKKO HINTIKKA, *Academy of Finland and Stanford University*

GABRIËL NUCHELMANS, *University of Leyden*

WESLEY C. SALMON, *Indiana University*

PHILOSOPHY OF PHYSICS

by

MARIO BUNGE

Foundations and Philosophy of Science Unit, McGill University, Montreal

D. REIDEL PUBLISHING COMPANY

DORDRECHT-HOLLAND/BOSTON-U.S.A.

199707

Library of Congress Catalog Card Number 72-86103

ISBN 90 277 0253 5

Published by D. Reidel Publishing Company,
P.O. Box 17, Dordrecht, Holland

Sold and distributed in the U.S.A., Canada and Mexico
by D. Reidel Publishing Company, Inc.
306 Dartmouth Street, Boston,
Mass. 02116, U.S.A.

Printed in The Netherlands by D. Reidel, Dordrecht

DEDICATION

To those who have undertaken to study physics for the love of it and who, despite course drillings, demands for fast results, and market pressures, still love their science, have not given up the hope of understanding it better, and dare to ask radical questions. For theirs is the Kingdom of Photons.

PREFACE

This book deals with some of the current issues in the philosophy, methodology and foundations of physics. Some such problems are:

– Do mathematical formalisms interpret themselves or is it necessary to adjoin them interpretation assumptions, and if so how are these assumptions to be framed?

– What are physical theories about: physical systems or laboratory operations or both or neither?

– How are the basic concepts of a theory to be introduced: by reference to measurements or by explicit definition or axiomatically?

– What is the use of axiomatics in physics?

– How are the various physical theories inter-related: like Chinese boxes or in more complex ways?

– What is the role of analogy in the construction and in the interpretation of physical theories? In particular, are classical analogues like those of particle and wave indispensable in quantum theories?

– What is the role of the apparatus in quantum phenomena and what is the place of measurement theory in quantum mechanics?

– How does a theory face experiment: single-handed or with the help of further theories?

These and several other questions of the kind are met with by the research physicist, the physics teacher and the physics student in their everyday work. If dodged they will recur. And a wrong answer to them may obscure the understanding of what has been achieved and may even hamper further advancement. Philosophy, methodology and foundations, like rose bushes, are enjoyable when cultivated but become ugly and thorny when neglected.

There are no prerequisites to the reading of this book except for some undergraduate theoretical physics and a native interest in the subject. The book can be used for independent reading as well as for a one-semester course at the senior or graduate levels.

I am grateful to the Canada Council for a Killam research grant that made the completion of this work possible.

MARIO BUNGE

CONTENTS

CHAPTER 1

PHILOSOPHY: BEACON OR TRAP*

There was a time when everyone expected almost everything from philosophy. It was the time when philosophers drew confidently the main lines of a picture of the world and left physicists the menial task of filling in some details. When this aprioristic approach was seen to fail, the physicist forsook philosophy altogether. Today he expects nothing good from it. So much so that the mere word 'philosophy' is apt to evoke an ironic or even contemptuous smile in him. He knows better than to indulge in free wheeling in the void.

However, the neglect of philosophy will not stave it off. Indeed, when we say that we do not care for philosophy, what we are likely to do is substitute an implicit, hence immature and uncontrolled philosophy for the explicit one. The typical physicist of our time has discarded the worn out dogmatic systems – which were half untestable and half false, and largely sterile anyhow – only to adopt uncritically an alternative set of philosophical tenets. This home-spun philosophy, extremely popular in the physical profession since the dawn of our century, goes by the name of *operationism*. It holds that a symbol, such as an equation, has a physical meaning only to the extent to which it concerns some possible human operation. Which entails that the whole of physics is about operations, chiefly measurements and computations, rather than about nature. Which constitutes a comeback of the anthropocentrism prevailing before the birth of science.

The physics student absorbs the operationist philosophy from the very beginning: he finds it in textbooks and courses as well as in seminar discussions. He seldom encounters any critical scrutiny of this philosophy, for this examination is usually done by philosophers whom he does not read. Moreover, if he feels tempted to criticise the official philosophy of science he may soon discover that he is not supposed to do it. Operationism is the orthodox credo, and any deviation from it is likely to be derided or even punished.

In any case, both the operationist and his critic philosophise. Philosophising is not unusual and it is not hard: what is difficult is to do some good philosophy, even more difficult to abstain from philosophy altogether. In short, the physicist is not philosophically neutral. He holds, mostly unwittingly, a set of philosophical tenets which shall presently be examined.

1. THE STANDARD PHILOSOPHY OF PHYSICS

The contemporary physicist, no matter how sophisticated and critical he may be on technical matters, usually espouses dogmatically what may be called the Credo of the Innocent Physicist. The main dogmas of this credo are the following.

(I)   Observation is the source and the concern of physical knowledge.

(II) Nothing is real unless it can become part of human experience. The whole of physics concerns experience rather than an independent reality. Whence physical reality is a sector of human experience.

(III) The hypotheses and theories of physics are but condensed experience, i.e., inductive syntheses of experiential items.

(IV) Physical theories are not created but discovered: they can be discerned in sets of empirical data, such as laboratory tables. Speculation and invention play hardly any role in physics.

(V) The goal of hypothesising and theorising is to systematise a part of the growing fund of human experience and to forecast possible new experiences. In no case should one try to explain reality. Least of all should we attempt to grasp essentials.

(VI) The hypotheses and theories that include nonobservational concepts, such as those of electron and field, have no physical content: they are merely mathematical bridges among actual or possible observations. Those transempirical concepts, then, do not refer to real yet imperceptible objects but are just auxiliaries devoid of reference.

(VII) The hypotheses and theories of physics are not more or less true, or adequate; since they correspond to no independently existing items, they are only more or less simple and effective ways of systematising and enriching our experience rather than components of a picture of the world.

(VIII) Every important concept has to be defined. Consequently every well-organised discourse has to start by defining the key terms.

(IX) What assigns meaning is definition: an undefined symbol has no physical meaning and therefore can occur in physics only as a mathematical auxiliary.

(X) A symbol acquires a physical meaning through an operational definition. Whatever is not defined in terms of possible empirical operations is physically meaningless and should therefore be discarded.

Give or take some commandment, most contemporary physicists seem to pay at least lip service to the preceding Decalogue – not just in the Western World but in the other worlds as well. This does not entail that all those who swear by the Decalogue live up to it. As a matter of fact no physicist would get very far if he were to act in abeyance of the Decalogue, for the latter neither reflects real research nor promotes it. This is what I shall try to show next – i.e. that operationism is a phony philosophy of physics.

2. Observation and reality

Axiom I, which makes observation the fountainwell and object of physical knowledge, is partially true: there is no doubt that observation does supply some rudimentary knowledge. But even ordinary knowledge goes far beyond observation, as when it postulates the existence of unobservable entities such as the interior of a solid body and radio waves. And physics goes even farther, by inventing ideas that it could not possibly extract from common experience, such as the concept of meson and the law of inertia. In sum, it is false that observation is the spring of every item of physical knowledge. As false as the claim that the good observations are those untainted by theory.

Besides, observation, regarded as an act, is not a concern of physics but rather psychology. Thus the theory of elasticity is about elastic bodies rather than about human observations of such bodies. Were it not so, the specialist in elasticity would observe the behaviour of fellow physicists rather than the one of elastic bodies, and he would propose hypotheses concerning the knowledge of those things instead of trying out hypotheses about the inner structure and the overt behaviour of elastic bodies. What is true is that some of the elementary problems in elasticity were suggested by intelligent (i.e. theory-soaked) observation, and that every theory of elasticity should be tested by experiments involving observations. But this is not what Postulate I claims.

Axiom II, which belongs in metaphysics, seeks to dispense with the concept of reality; at the very least, it tries to bracket it out during scientific investigation. Until the era of operationism every physicist thought he was manipulating real things or having ideas concerning them. This is what he still does when at work, though not when he philosophises: on these occasions, the practical realist often turns into an empiricist. Only some conservatives like Einstein dared to hold, in the very heyday of operationism, that physics attempts to know reality. The mistrust of the reality concept seems to have been inherited from the British empiricists and from Kant – via the positivists and pragmatists – who criticised the claims of the schoolmen and other speculative philosophers, that they were able to grasp an immutable reality underneath the changing human experiences. But this involves a very special use of the term 'reality', one that has only a historical interest. And anyhow it is uninteresting to flog the dead horse of traditional metaphysics: what is of interest is to find out whether physics is really tied to a metaphysics of experience, rather than to the older metaphysics of substance, or whether it condones neither of them.

Surely physics does not exclude the concept of reality but restricts it to the physical level, leaving other sciences the task of investigating other levels – in particular the one of human experience. No physical theory makes the assumption that its object be feelings, thoughts, or human actions: physical theories are about physical systems. Moreover, even though physics is not concerned with human experience it constitutes a radical extension and deepening of human experience. Thus, producing a 1 GeV particle beam is a novel human experience and so is understanding the scattering of that beam by a given target, but the point of designing and performing the experiment, as well as of working out the respective theory, is to know more about particles, not about men. Likewise the astrophysicist who studies thermonuclear reactions in the interior of stars does not penetrate them except intellectually: he has no direct experience of the objects of his study. Yet he believes, or at least he hopes, that his theories do have real counterparts. Of course this belief, or rather hope, is not a groundless one: unlike the metaphysician of old, the scientist checks his theories by contrasting them with observational data – many of which may have been gathered in the light of the very theories he puts to the test. In other words, while experiences of various kinds are needed to test our physical ideas, they do not constitute the referents of the latter. The intended referent of any physical idea is a real thing. If this particular thing does not happen to be real, so much the worse for the idea. Reality does not seem to care for our failures. But if we neglect reality or deny that there is any, we end up by giving up science and espousing the worst possible metaphysics instead.

3. NATURE OF PHYSICAL IDEAS

Axiom III, concerning the nature of physical hypotheses and theories, extrapolates to physical science what holds for a part of ordinary knowledge. It is true that many general statements are inductive syntheses or summaries of empirical data. But it is false that every general physical idea is formed by induction from individual experiences, e.g. observations. Consider the formulas of theoretical physics, even of the hardest kind – those of solid state physics. All of them contain more or less sophisticated theoretical concepts that are removed from immediate experience. What is more, hypotheses and theories exude experience rather than summarising it, for they suggest new observations and experiments. Yet this is not the most important of the functions of hypotheses and theories: we value them primarily because they enable us to trace a more or less sketchy map of reality and because they enable us to explain the latter, even if partially and gradually.

Nothing is being explained by saying that something is a fact of experience, or by making sure that a statement is a package of experiential items. Experience is something to be explained, and explanation is a task for theories. In particular physical theories, rather than being units of canned experience, allow us to account for one side of human experience, itself a minute part of reality. But they do not suffice, for every human experience is a macrofact with many aspects and occurring on a number of levels, from the physical level to the mental one, so that a suitable explanation of it calls for the cooperation of physical, chemical, biological, psychological, and psychosocial theories. In short, physical ideas go much beyond experience and this is why they can contribute to explaining experience. The third axiom of the official philosophy of physics is, then, false. It is also obnoxious, as it reinforces the myth that, while all experiments are important, no theory is indispensable.

Axiom IV is really a consequence of Axiom III: if theories are inductive syntheses then they are not created but are formed by agglomerating empirical particulars, much in the same way as a cloud is formed by the aggregation of water droplets. The falsity of this thesis follows from the falsity of Postulate III, but it can be exhibited independently by recalling that every theory contains concepts that do not occur in the data employed in checking it. Thus continuum mechanics employs the concept of inner stress; but, since this concept is unobservable, it does not figure in the data used to support or to undermine any particular hypothesis concerning the definite form of the stress tensor.

A further argument, of a psychological nature, can be wielded against Postulate IV, namely this. No physical theory has ever come out of the contemplation of things, or even of empirical data: every physical theory has been the culmination of a creative process going far beyond the data at hand. This is so not just because every theory contains concepts that do not occur in the experimental statements relevant to it, but also because, given any set of data, there is an unlimited number of theories that can account for them. There is no one way road from data to theories; on the other hand the way from the basic assumptions of a theory to its testable consequences is unique. In short, while induction is ambiguous, deduction is unambiguous. Besides, theories are not photographs: they do not resemble their referents but are symbolic constructions built in every epoch with the help of the available concepts. Scientific theories, far from being inductive syntheses, are creations – subject to empirical test to be sure, but not any less creative for that.

4. GOAL OF PHYSICAL IDEAS

Axiom V, concerning the goal of physical ideas, is one-sided and it presupposes that there is just one goal. It is true that systematising or ordering is one of the aims of theorising, but it is not the only one. The synoptic table, the numerical table and the graph are as many ways of compressing and ordering data, but none of them suffices to explain why things should happen as they do rather than otherwise. In order to explain something we must deduce statements describing that fact, and deduction calls for premises going beyond what is being explained.

These premises are so many hypotheses containing theoretical concepts. In short, the main function of physical theories is to provide explanations of physical facts.

But there are superficial explanations and deep ones, and we shall not settle for the former if we can get the latter. Now to explain in depth, to go to the kernel of things, we must hypothesise mechanisms – not necessarily or even usually mechanical ones. And mechanisms, except for the macrophysical and properly mechanical ones, escape perception. Only deep (nonphenomenological) theories can account for them. In short, to get deep explanations, be it in physics or in any other science, deep theories must be invented: theories transcending both experience and the black box type theories.

In many cases such deep theories seem to get close to the essence of their objects – or rather to the essential or source properties of them. Hence it can no longer be held that physics, for not going beyond relations and regularities, does not grasp the essence of things. There are essential or basic properties, such as mass and charge, which originate several other properties; likewise there are basic or essential patterns, involving some of those source properties, which give rise to derivative patterns. Surely there are no immutable essences that intuition alone can grasp. Moreover, any hypothesis concerning the essential character of a given cluster of properties and laws is subject to correction. But the fact is that, insofar as physics transcends the external or behaviorist approach – which is necessary but insufficient – it undermines Postulate V.

5. THEORETICAL CONCEPTS AND TRUTH

Axiom VI is common to conventionalism, pragmatism and operationism (which may be regarded as the philosophy of science of pragmatism). If adopted, most of the referents of physical theory are dropped and we are left with empty calculi. For, what characterises a physical theory by contrast to a purely mathematical one, is that the former concerns – whether rightly or wrongly – physical systems. If a theory is not about a class of physical systems, then it does not qualify as a physical theory. Hence the sixth dogma is semantically false. It is psychologically false as well, for if theories were nothing but data grinding machines, nobody would bother to build them: the theoretician's aim is to produce an account of a piece of reality. In short, Postulate VI is false on all counts. Nevertheless, it has had the historical merit of discrediting naive realism: we now begin to understand that physical theories are not portraits of reality but that they involve brutal simplifications leading to ideal schemas, or object models, such as those of homogeneous field and free particle. We also acknowledge that, in addition to such first approximations, we must introduce conventions such as those of measurement units. But none of this turns physics into a mere fiction or into a set of conventions, just as an ordinary language description of an observable phenomenon is not hollow for being couched in a conventional system of signs.

As to *Axiom VII*, which seeks to eliminate the concept of truth, it follows from the conventionalist thesis. For, if physics is not about real objects, then its statements are not such, i.e. they are not more or less true (or false) formulas. But this doctrine does not tally with the physicist's practice. In fact, when the theoretician derives a theorem, he claims that the latter is true in the theory, or theories, to which it belongs. And when the experimenter confirms that theorem in the laboratory, he infers that the statement is true, at least partially, relative to the empirical data concerned. In short, both the theoretical and the experimental physicist use the concept of truth and they may even feel insulted if told that they are not after the truth.

Surely the truths attainable in physics are relative truths in the sense that they hold, if at all, relative to certain sets of propositions that are momentarily taken for granted, i.e. not questioned in the given context. They are also partial or approximate truths, for confirmation is always partial and moreover temporary. But truth is not an illusion for being relative and partial. As to the simplicity and efficiency that the pragmatist worships in the place of truth, they are not found in every theory. The deepest physical theories, such as general relativity and quantum mechanics, are also the richest. And practical efficiency can only be obtained when passing over to applied science or technology. Whether simple or complex, a physical theory is neither efficient nor inefficient but more or less true. A coarse theory applied with skill to practical ends can be as effective as a refined theory, even though normally the greater the truth the greater the efficiency. In any case, efficiency is not inherent in theories: it is a property of the pairs ends-means; theories occur among the means employed in technology, but it is only in relation with the goals that their efficiency can be judged. The upshot is that the VIIth postulate of the official philosophy of physics is false.

6. DEFINITION

Axiom VIII, which demands that every concept should be defined to begin with, is flatly absurd. A concept, if defined, is built in terms of other concepts, so that some of them must remain undefined. Thus the concepts of mass and force are primitive (undefined) in Newtonian mechanics. They are not therefore obscure or indeterminate, since they are specified by a number of formulas. A well-built theory does not start with a bunch of definitions but rather with a list of undefined concepts, or primitives. These are the units that, glued with logical and mathematical concepts, occur once and again at every stage in the construction of a theory. They are the essential or basic concepts in a given theory, those which cannot be dispensed with. All the other concepts, i.e. those definable in terms of the primitives, are logically secondary. Hence the VIIIth dogma, to which so many textbooks try to adjust, is wrong.

Axiom IX, concerning the procedure by which a meaning is assigned to a symbol, does not hold in general. Definitions assign meaning on condition that they be framed in terms of symbols that have a meaning themselves. And such defining symbols cannot be assigned a meaning by definitions precisely because they are defining, not defined. Therefore a means other than definition must be resorted to in order to delineate the meaning of a basic or undefined physical symbol.

The best we can do is to lay down all three conditions the symbol must satisfy: (a) the mathematical conditions, i.e. the formal properties it is supposed to have, (b) the semantical conditions, i.e. what physical object or property it is supposed to represent, and (c) the physical conditions, i.e. the relations it is assumed to hold to other physically meaningful symbols in the theory. Since every condition of this sort is an axiom or postulate, we see that the task of assigning physical meanings in an unambiguous and explicit way is performed by axiomatising the theory in which the symbols concerned occur. (More on this in Chapters 7 and 8.) Thus in continuum mechanics 'T' is a primitive symbol that designates a concept – the one of inner stress – that has a definite mathematical form (namely a tensor field over a four dimensional manifold) and a definite referent (namely a property of a body). This last assumption, of a semantical nature, is not a convention like a definition, but a hypothesis. Indeed, it might turn out to be empty; moreover, for all we know there are no continuous material bodies. But the theory does make the hypothesis that such things exist. And if the theory works then bodies may be nearly continuous. In sum, what assigns a meaning to a basic physical symbol is not a definition but a whole theory, with its three ingredients: the mathematical, the semantical, and the physical assumptions. Should the theory prove false, its primitives would still keep a definite meaning, but they would become idle. In any case, the IXth dogma is wrong, for only the defined or secondary symbols are given a meaning via definitions.

7. OPERATIONAL DEFINITION

Finally, also *Axiom X*, relative to the so-called operational definitions, is false. When applied to the case of the electric field strength E this dogma holds that 'E' acquires a physical meaning only when a procedure for measuring the values of E is prescribed. But this is impossible: measurements allow us to determine only a finite number of values of a function, and moreover they yield only rational or fractionary values. Besides, the numerical value of a magnitude or physical quantity is only one constituent of it. For example the concept of electric field is, mathematically speaking, a function and therefore it has three ingredients: two sets (the domain and the range of the function) and the precise correspondence between them. A set of measured values is only a sample of the range of the function. Unless one has a fairly well-rounded idea of the whole thing one would not even know how to go about taking such a sample. That is, far from assigning meanings, measurement presupposes them.

Moreover, the measurements of the value of E are always indirect: fields are accessible to experience only through their ponderomotive actions. What is more, there are many ways of measuring values of E. Hence if every one of them were to determine one concept of electric field strength, we would have a number of different concepts of electric field rather than the single concept entering Maxwell's theory. If we wish to know what 'E' means we must look into Maxwell's theory. Meanings are not determined by doing but by thinking. It is only once we have a reasonably clear idea that it pays to go to the laboratory. In sum, Axiom

X is false: there are no operational definitions. The belief that there are comes from an elementary confusion between defining (a purely conceptual operation and moreover one that does not apply to the basic concepts) and measuring – an operation which is not only empirical but conceptual as well.

This closes our criticism of the Credo of the Innocent Physicist. It has employed a few philosophical tools – mainly logic and semantics – and a few counterexamples taken from physics. The upshot is clear: to the extent to which our criticism is justified, philosophy done in an explicit way can be useful to lift some of the fog that hovers over physics.

8. TOWARDS A NEW PHILOSOPHY OF PHYSICS

The failure of operationism does not put an end to the philosophy of physics. There are plenty of alternatives to operationism and they have been around for a long time: almost every school philosophy is one. However, most school philosophies other than operationism have failed to attract the attention of physicists, and this for the following reasons. First, those philosophies are the work of professional philosophers rather than of scientists, and it is only natural (though not altogether rational) that a scientist should feel inclined to distrust philosophers and trust instead a fellow scientist who seems to spin out a philosophy all by himself and who talks his own language. Second, the general philosophies that compete with operationism are usually much too general and sometimes also obscure: they seldom take pains to perform a detailed analysis of a genuine piece of science and on the other hand they stress extrascientific points (religious, political, etc.) that have no immediate bearing on scientific theories or experiments. Third, most philosophies of physics other than operationism are hardly relevant to physics: they are unconcerned with actual cases of theorising and experiment and play around with miniproblems the solution to which would make no difference one way or another: they miss the real issues or attempt to handle them without any specialised knowledge. In short, there are reasons, at least two good ones, for the physicist's lack of interest in most of the philosophies of science.

The failure of both operationism and the traditional philosophies of physics to perform adequate philosophical analyses of physics constitutes a challenge to build an alternative philosophy of physics. The new philosophy that physics needs should be both its conscience and its wing: it should help physics criticise itself and also explore new problems and methods. The main ingredients of this new philosophy of physics *(NPP)* should be these:

> *Equation of motion:* The specific input to *NPP* should be the whole of physics, past and present, classical and quantal. The corresponding output should be a realistic account (analysis and theory) of actual and optimal research procedures, of conceived and conceivable ideas, of currently pursued and possible goals both in theoretical and experimental physics. *Constraint: NPP* is to keep pace not only with advances in physics but also with the relevant advances in exact philosophy, particularly in logic and semantics.
> *Boundary condition: NPP* should make the most of the philosophical tradition, by assimilating it critically.

This new philosophy of physics is in the making: we shall sample it at various places in this book. For the moment let us list some of the problems that are currently being investigated in the spirit of the new philophy: this, rather than an enumeration of workers and works, may give an idea of the vitality of the field and of its relevance to actual physical research. Here they go in happy disorder:
– Does relativity to a reference frame amount to dependence on the observer and thus to subjectivity?
– Does invariance under coordinate transformations ensure both meaning and objectivity?
– Are quantum events inconceivable without the intervention of an observer?
– Does the quantum theory concern autonomous physical objects or rather unanalysable blocks formed by the fusion of micro-objects, measurement instruments, and observers?
– Are there strictly observational concepts in physical theories?
– How observable are the so called observables of quantum theory and general relativity?
– What are the aims of a physical theory: to systematise data, to compute predictions, to guide further research, and/or to explain facts?
– Is it true that one cannot explain without resorting to familiar images

or pictorial models, and that as a consequence quantum mechanics and general relativity have no explanatory power?
– Is it possible to make experiments without the help of theories and thus collect theory free data?
– What does the physical meaning of a symbol consist in?

Hundreds of further problems in the current philosophy of physics can be gleaned from the literature, in particular from the quarterlies *Philosophy of Science, British Journal for the Philosophy of Science, Synthese,* and *Dialectica.* Still, such a wealth of questions (not to mention the answers) constitutes no proof that the philosophy of physics, even if free from the defects of either operationism or traditional school philosophy, does serve a useful purpose. Let us turn to this question.

9. THE FUNCTIONS OF PHILOSOPHY

The philosophy of physics is a branch of the philosophy of science, alongside similar disciplines such as the philosophy of biology and the philosophy of psychology. In turn, the philosophy of science is but one of the branches of philosophy – others being logic, general epistemology, metaphysics, value theory, and ethics. We have seen that a wrong philosophy may prevent a correct understanding of physical theory and experiment. It may even delay progress in research by banning whole research programmes incompatible with that philosophy or by fostering superficial or even sterile programmes. Can philosophy do better than this? Can it perform any positive function? It certainly can and sometimes it does. Historical examples abound but we shall not use them, for the declared adoption of a given philosophy does not prove its observance. We are interested in the functions of philosophy that are conceptually possible.

The philosophy of physics can perform at least four useful functions, which may be called philosophical assimilation, research planning, quality control, and house cleaning. Explicitly:

(i) The *philosophical assimilation* of physics consists in enriching philosophy by processing ideas and methods developed in physics. By analysing the actual work of experimental and theoretical physicists the epistemologist may conceive general hypotheses concerning the nature of human knowledge and the ways to make it grow or decay. By examin-

ing deep physical theories the metaphysician may invent general theories
about the nature of things. In short, the philosophy of physics may
contribute (and as a matter of fact has often contributed) to the expansion
or even the renewal of philosophy.

(ii) *Research planning* is always done with some philosophy or other in
mind. If the guide (or misguide) is a narrow empiricist philosophy, research will be limited to data gathering and to phenomenological or black
box theories covering those data without explaining them. On the other
hand if a more liberal philosophy is adopted then no limitations will be
imposed on the depth of the theory nor on the dependence of experiments
on theories. In particular, the search for bold theories and for new kinds
of data will then be encouraged rather than decried. Budget is only one
item to be considered in programming research: the philosophy of science
is an even more important item and one that will partly determine the
budget size, since philosophy shapes the very goal of research. If the aim
is to multiply data, ask for increased instrument and computer facilities.
If the goals are to find new laws and to evolve and test ambitious theories,
ask for more ingenious experimenters and theoreticians.

(iii) The *quality control* of research consists in checking and evaluating
the worth and significance of experimental and theoretical results. Are
the data reliable? Are they of any value to test theories, or set them in
motion, or pose questions calling for new theories? Are the theories worth
while? The answer to any of these questions involves some philosophical
ideas on the nature of truth, the interplay of experience and reason, the
structure of scientific theories, and so on. Just consider the various criteria
proposed for evaluating the truth claims of a theory: to some, the seal of
truth is simplicity; to others it is beauty; to most, a strong empirical
confirmation; to many, technological performance; and so on.

(iv) By *house-cleaning* I mean, of course, the never-ending process of
clarifying ideas and procedures. Surely the actual formation of new
physical concepts, hypotheses, theories and procedures is the task of the
professional physicist. But subjecting them to a searching critical scrutiny
calls for some logical, epistemological and methodological rigor. And
letting them live and show their worth requires a tolerance that only a
good philosophy can teach.

Research planning, quality control of end products and house-cleaning
involve then some philosophy: the physicist who undertakes either task

becomes a part-time philosopher. And a part-time philosopher will make the most of his time if properly equipped.

10. THE ROLE OF PHILOSOPHY IN THE TRAINING OF PHYSICISTS

Any physicist who scratches the surface of his own work is bound to come face to face with philosophy even though he may not realise it. If he recognises the beast he has two possibilities. One is to let himself overcome, i.e. to succomb to the prevalent philosophy which, being popular, is bound to be coarse and even backward. The other possibility is to study the beast hoping to tame it – i.e. to become acquainted with some of the current research in the philosophy of physics, examining it critically and trying to put it in the service of his own scientific work.

The physicist who refuses to let himself chain to an anachronistic philosophy and is willing to regard philosophy as a possible field of exact inquiry may expect a lot from such an approach. The reading of imaginative philosophers may suggest him new ideas. The study of logic will raise his standards of clarity and rigor. The habit of semantic analysis will help him discover the genuine referents of his theories. A familiarity with professional doubters will protect him against dogmatism. An acquaintance with huge unsolved problems and with grand schemes will encourage him to undertake long-term research programmes instead of hopping from one fashionable little problem to another. Awareness of the methodological unity of all the branches of physics and indeed of all the sciences will prevent his overspecialisation – a major cause of unemployment in the profession's crisis at the time of writing. If nothing else, a pinch of philosophy will reinforce the theoretician's and the experimentalist's faith in the power of ideas and in the need for criticism.

In short, philosophy is always with us. Hence the least we should do is to get acquainted.

NOTE

* Some paragraphs are reproduced from Bunge (1970b) with the permission of the editor and publisher.

CHAPTER 2

FOUNDATIONS: CLARITY AND ORDER*

Most physicists can spare no time to analyse the very concepts, hypotheses, theories and rules they create or apply: they are too busy devising or using them. Which is just as well: it would be pedantic, and also hardly profitable, to press every particle physicist for a neat analysis of the very concept of particle. This does not mean that conceptual analysis lacks in value for physics: it is valuable but it need not engage everyone. Likewise only a fanatic would wish to proscribe conceptual analysis as a legitimate occupation of some physicists: after all, someone ought to analyse and even refine what others create. A large undertaking like physics takes all kinds: from experimentalists of various descriptions to theoreticians of all sorts: each of this species is needed to explore and understand nature.

There are still others who contribute to the same goal although they are not intent on finding physical laws: they are the instrument designers and makers, and the mathematical physicists. The former deal with *artiphysis* rather than with *physis*, and the mathematical physicists focus on mathematical problems posed by the development of physical theories. Yet they are not looked down upon as if they were hangers-on. The analyst of physics is in a similar position: although he is not expected to make discoveries about physical reality, he can help find out what physics is all about, by analysing and thereby elucidating some of the basic concepts, hypotheses, theories and procedures of physical science. And his help will be even more valuable if, in addition to analysing physics, he contributes to laying bare the organisation or structure of physical theories and of whole sets of physical theories, i.e., if he becomes an accomplished craftsman in the foundations of physics.

Few physicists think of the organisation of physics. Most are content with letting it grow, and some are even opposed to any kind of organisation. Mathematicians, on the other hand, are keen on the organisation of algebra, topology, analysis, and indeed the whole of mathematics: they have realised that concern for structure facilitates growth by showing relations, gaps and shortcomings that were not seen when concentrating on single points. Thus during our century algebra has thrice been subjected to a thorough shake-up bearing primarily on the organisation of its material but ensuing in its enrichment: first by axiomatisation, later by logic, and more recently by categorisation (formulation in terms of category theory). These three revolutions have brought to algebra not only unity but also a greater depth and scope. Similarly analysis has thrice been revolutionised in the course of the last hundred years: first by arithmetisation, then by set theory, finally by topology – not to speak of the fourth current revolution, namely categorisation. When a new basic concept is employed in the organisation of a field of mathematics (like the concepts of set, structure, and functor), both an internal and an overall reorganisation may ensue: the former bearing on single theories, the latter on whole families (categories) of theories. Why should physicists regard it as beneath their dignity to work on a similar project, namely the explicit organisation of theoretical physics? Why should order be dreaded rather than encouraged?

In sum, there are plenty of questions concerning the analysis and the organisation of theoretical physics: while some of these questions are philosophical, others are technical questions that can only be answered with the help of technical tools such as logic, mathematics and axiomatics, and are thus as unlikely to attract the bulk of philosophers as they are unlikely to divert the attention of most physicists. They are questions in the foundations of physics. Let us be more specific and proceed to formulate some of them.

1. SOME CURRENT PROBLEMS IN THE FOUNDATIONS OF PHYSICS

A practical way of introducing a field of research and showing its significance is to list some of its typical problems. Here goes a score of problems that are currently engaging the attention of a number of workers in the foundations of physics, as can be seen by perusing not only philosophical journals but also physics journals such as *J. Math. Phys.*, *Progr. Theor. Phys.*, *Rev. Mod. Phys.*, *Intern. J. Theor. Phys.*, *Nuovo Cimento Suppl.*, *Amer. J. Phys.*, and *Foundations of Physics*.
– Exactly at which point, in the development of a theory containing spatio-temporal concepts, does it become necessary to introduce coordi-

nates? Equivalently: how far can we go in a coordinate free (hence automatically general covariant) formulation?

– Could we read some of the properties of space time from certain physical laws? (Not obvious: Maxwell's equations need no commitment to a metric.)

– Is it true that the so-called direction of time has to be sought in irreversible processes and that time itself has got to be defined in terms of irreversible processes?

– Is time really equivalent to a spatial dimension?

– What if any are the limits of spatiotemporal localisation?

– Does quantum mechanics concern individual microsystems or only statistical ensembles of such, or even ensemble-apparatus pairs?

– How do we have to interpret the probabilities occurring in physical theories: as strengths of our beliefs about the physical system concerned, as relative frequencies of measured values, or as tendencies (propensities)?

– How should we interpret the dispersions or scatters occurring in the so called indeterminacy relations: as uncertainties, as standard deviations of sets of measurement results, as objective indeterminacies, or perhaps in some other way?

– Is it possible to explain chance by deducing every stochastic theory from a deeper deterministic theory?

– Is it possible to retrieve the quantum theory from some classical stochastic theory?

– Does the quantum theory require a logic of its own, e.g. one excluding the conjunction of statements concerning the exact values of conjugate dynamical variables?

– Does the correspondence principle constitute an integral part of the quantum theory or a rule belonging to its heuristic scaffolding?

– Are covariance principles and symmetry statements (like the CPT theorem) among the axioms of a theory? And do they refer to physical systems?

– Is it possible to give an observer-free formulation of quantum mechanics and of quantum electrodynamics?

– Is it possible to deduce continuum mechanics and thermodynamics from particle mechanics?

– What are the relations among the various theories in the current map of physics?

– Does a quantum mechanical explanation of a bulk property, such as the refractive index or the electric conductivity, consist in a reduction of the macrolevel to a microlevel?

– Are physical quantities anything more than functions of a certain kind? And what is the difference if any between a physical quantity and a dimensional constant or scale factor?

– What are the algebras of dimensions and of units, that underly the standard rules for handling them?

– What are the relations between units and standards? Granting that units are conventional, how about standards?

This brief list could be used as a questionnaire for measuring one's attitude towards the foundations of physics. Thus while to the writer they are all interesting and some of them constitute Ph. D. subjects, presumably many people regard them as dull or trivial questions or even as irrelevant to physics. But then many other fields suffer the same fate: thus one of my teachers held magnetism to be a very dull subject, and I confess to my inability to get excited by classical acoustics. *De gustibus non est disputandum.* What really matters is whether there are open problems in a given field and whether solving these problems would make any difference to our understanding and enjoyment of nature or of natural science. And this we won't know for certain, in the case of the foundations of physics, unless we give it a chance.

2. THE SEARCH FOR ORDER AND COGENCY

The foundations of physics have been assigned two chief missions: to enhance the clarity of physical ideas and to improve their organisation. I will argue that the first task is best accomplished through the second. Let us then retake the matter of structure.

Order and cogency have not only an aesthetic value: the better organised a body of ideas the easier it is to grasp and retain it (psychological advantage) and the better it lends itself to evaluation, criticism, and eventually to its replacement by a different system of ideas. For these reasons mathematicians have, from the time of Euclid, valued the axiomatic formulation of theories. This is not just a matter of taste nor even mainly a matter of teaching but, as intimated before, one of methodology: axiomatics is scientifically valuable because it renders explicit all the assumptions actually employed and so makes it possible to keep them under control.

What holds for pure mathematics holds, in respect of axiomatics, for every one of its applications, from physics through social science to philosophy: there is little hope for order, cogency and even relevance outside axiom systems. For axiomatics, in addition to bringing order and the possibility of spotting inconsistency, enables us to spot irrelevancies and on occasion even whole formulas that pass for deep principles or theorems just because they make no sense in the given context. Three examples from contemporary physics will suffice to show how a bit of axiomatics may help to expel intruders that perform no function either in computation or in measurement and are kept by sheer force of authority.

Our first example is the pseudoconcept of photon mass. When an enthusiast of "$E^2/c^2 = m_0c^2 + p^2$" speaks about the photon mass one can remind him that this is a phoney predicate, as it fails to occur among the basic concepts of electromagnetic theory: the above formula belongs to relativistic particle *mechanics* – and mechanics, however refined, is not competent to account for photons. Moreover, that formula is but the consequent of an implication with the antecedent: "A particle has a mass m, linear momentum p and energy E". The converse conditional is false: not every energy can be assigned a mass and a mechanical momentum $p = mv$. Hence it is mistaken to talk about the universal equivalence or interconvertibility of mass and energy.

Second example: when theoreticians and even experimentalists draw a distinction between inertial mass and gravitational mass (only to equate them immediately thereafter), one may remark that no known theory has been proposed in which different (rest) mass concepts occur (Bunge, 1967a). If such a distinction is meant seriously then it must be formulated axiomatically: each mass concept must be characterised by one or more axioms rather than by pseudophilosophical or heuristic remarks. If no such distinction is made then they are not different – period.

Our third and last example will keep us a little longer because it is slightly more complicated: it is the so-called fourth indeterminacy relation. In his last discussion with Einstein on epistemological problems in atomic physics, Niels Bohr (Bohr, 1949) claimed that time and energy satisfy an "uncertainty" relation similar to Heisenberg's. More precisely he asserted, on purely heuristic grounds, that the mean standard deviations $\Delta_\psi t$ of time and $\Delta_\psi E$ of energy, for a quantum mechanical system in a state ψ, are related thus:

$$\Delta_\psi t \cdot \Delta_\psi E \geqslant \hbar/2 .$$

Unlike the genuine indeterminacy relations, the previous formula has never been proved from first principles. Moreover, as we shall see presently, it cannot be so proved. Had Einstein realised this his reply would have carried more weight and the whole discussion would not have been regarded by almost everyone as won by Bohr.

The reason for the failure to incorporate the above formula into quantum mechanics is the following. In this theory, as in every other known and successful theory, time is a "c number" and, more particularly, a parameter of a certain group of transformations: it is not a dynamical variable on a par with the position and the momentum operators. Moreover, unlike the latter, t does not "belong" (refer to) the particular system concerned but is public (at least locally). Even in relativistic theories the proper time, though relative to a reference frame, is not a property of the system on the same footing as its mass or spin. In other words, t does not belong to the family of operators in the Hilbert space associated to every pair microsystem-environment. Therefore t is not a random variable: no probability distribution is associated to it. Hence its scatter vanishes identically:

$$\Delta_\psi t = 0 \quad \text{for every} \quad \psi .$$

(Furthermore, the scatter in the energy vanishes when the system happens to be in an eigenstate of the energy operator.) As a consequence, no matter what the scatter in the energy may be, the inequality proposed by Bohr and repeated in many textbooks does not hold: it does not belong in the quantum theory, whether nonrelativistic or relativistic (Bunge, 1967a). This would have been realised much sooner if physical axiomatics had been taken seriously.

3. THE AXIOMATISER AND THE PHILOSOPHER

Unfortunately most physicists mistrust axiomatics, apparently because they believe that axiomatising is crystallising or ossifying. (One eminent physicist told the writer: "Axiomatisation is useless." Another went further, assuring him: "We do not want axiomatic theories in physics." No reasons were given in either case: *magister dixit*.) Whether one likes it or not, the fact is that an intuitively formulated theory is not so much a single theory as a set of theories, as many as different bunches of tacit assumptions are made. This is why any more or less amorphous theory can be axiomatised in a number of inequivalent ways, i.e. by adopting different backgrounds (e.g. different mathematical tools) and different basic hypotheses (axioms). Since axiomatising is rendering explicit what was tacit, the enemies of axiomatics unwittingly fight explicitness and favor its opposites, i.e. ambiguity and obscurity. Besides, axiomatising a theory does not force us to adopt it forever: rather on the contrary, since axiomatisation facilitates the scrutiny of the theory and eliminates any obscurities it may contain, it points the way to new theories gotten by changing some of the assumptions.

It may be rejoined that, even granting that axiomatics is valuable, this does not prove that philosophy is necessary for it. Granted: a good theoretician may axiomatise without availing himself explicitly of any philosophy, just as ordinary life does not gain much with the study of logic. But experience shows that the available physical axiom systems are mostly unbalanced: while some of them neglect specifying the mathematical status of the basic concepts, others do not specify clearly what the latter stand for. A pinch of philosophy might avoid the two extremes of concretism and formalism, because it is one of the tasks of philosophy to examine the nature of well-built scientific theories.

Consider once more the case of the symbol 'E' discussed in Chapter 1. The mathematician intent on axiomatising Maxwell's theory will certainly not forget to postulate, say, that 'E' designates a vector field over a certain differentiable manifold. But he may forget to say that this manifold happens to represent space time, and he may not care to state that the vector field refers to a supposedly real field spread over a region of space time. He may just hint at this intended interpretation or, adopting uncritically the philosophy of operationism, he may state that the numerical values of E are the results of measurements – which is neither exact nor specific enough. Or, finally, he may hold that 'E' is just a name for the expression 'electric field' – by which he reduces the semantical problem to the one of supplying rules of designation.

At this point the philosopher may point out that designation rules are hardly more than conventions whereby names are assigned, while semantical assumptions involve hypotheses concerning the existence of the referents. (Recall Chapter 1, Section 5.) He may also warn against the belief that an interpretation postulate will exhaust the meaning of the symbol concerned: he may point out that physical concepts are also specified by the mathematical and the physical assumptions, and not only by the basic ones but by the derived ones as well. The philosopher may remind the axiomatiser, in short, that physical meanings should not be neglected and that one should not believe that they can be assigned unambiguously by a sentence or two. To sum up, the philosopher may be of some help in the most delicate, though perhaps not the most creative, of theoretical activities, namely the foundation of theories.

4. THE SEARCH FOR CLARITY

Another aspect of foundations research is the analysis of theories, in particular of their distinctive concepts and statements. This analysis is usually performed in an intuitive or semi-intuitive way, i.e. without prior axiomatisation. But any rigorous analysis requires that the theory be there, fully and well ordered as far as its foundations are concerned. For example, it is absurd to try to find out whether the concept of electric field is primitive or derived except in a definite theoretical context. Besides, the meaning of 'E' may vary with the theory. Thus in one theory E will refer to a real field, a substance extending over a region of space; in another theory E will be no more than an auxiliary symbol, only the ponderomotive force eE being assigned a physical meaning. Finally, in an action at a distance theory E need not occur at all.

Here too the philosopher may be of help. For example, if the physicist is reluctant to assign 'E' a physical meaning in a field theory, the philosopher may press him for a reason for this reluctance. Should the physicist argue that E cannot be measured directly, and that no free fields can be measured, since the very presence of a measuring apparatus does away with the void, the philosopher may rejoin that a similar criticism, if extended to all other theoretical concepts, would deprive them of meaning. In any case, since the physicist who analyses a physical theory employs the philosophical concepts of theory, form, content, truth, and so many others, he can expect either criticism or help from the philosopher.

Contemporary scientific philosophy (mathematical logic, semantics, methodology, etc.) is then relevant to both the critical and the constructive (or rather reconstructive) aspects of foundations research. Philosophy certainly is insufficient: the subject must be mastered in the first place. But the physicist without philosophical competence is not much better off than the pure philosopher when it comes to doing foundations research. Thus in order to find out whether the mass concept is definable in mechanics, a knowledge of the latter is necessary but insufficient: a proof of concept independence requires a certain technique born in metamathematics and now belonging to the theory of theories (see e.g. Suppes, 1967). Another case of the same kind: any effort to give a context-free (i.e. theory-independent) definition of the concept of a simple system (e.g. an elementary particle) is bound to fail. It is only within the frame of some theory that the concept of a composite system can be defined in terms of the concepts of simple system and of a composition relation or operation. Of course experiment may refute the assumption that the system in question is simple, hence indecomposable: in this case the whole theory will have to be restricted to a more modest domain (e.g. for low energies) or even shelved altogether. But the point is that the concept of a simple system, like every other concept, can only be defined relative to some theoretical context: change the context and the concept may change or even vanish from sight.

When two different disciplines are jointly required to do a certain job, cooperation is mandatory. This is the case with the foundations of physics. The physicist unwilling to secure such a cooperation and who stubbornly refuses to look at the face of exact philosophy must resign himself to remaining ignorant about certain foundation problems and will have to count on a number of errors that could easily be avoided with a little philosophy. Common instances of such errors deriving from insufficient philosophy are: the belief that mass and energy are identical only because they are related; the belief that the use of probability is always indicative of incomplete knowledge; the belief that stochastic theories exhibit the bankruptcy of determinism; the belief that whatever is nonrandom must be causal; the belief that every theoretical value (e.g. the eigenvalue of a quantum mechanical dynamical variable) is a measured value – and hundreds of others which are repeated uncritically.

An exact analysis of a physical theory can only be made after the theory has been formulated in a full and consistent way, i.e. after it has been axiomatised. In the absence of such a reconstruction one can count only on one's intuition to traverse the tangle of formulas. What is worse, unless the theory is put together in an orderly fashion one will tend to pick on isolated formulas of the theory, typically de Broglie's or Heisenberg's formulas, forgetting where they come from, hence what their meaning is. Thus even though the Lorentz transformation formulas are derived without assuming any measuring instruments, they are usually interpreted as relating measurement results. And although Heisenberg's scatter formulas are likewise deduced without assuming any measurement, it is often claimed (a) that they result from an analysis of certain gedanken-experiments and (b) that they relate errors of measurement or even subjective uncertainties concerning the precise dynamical state of the object.

It is only when the two theories in question – special relativity and quantum mechanics – are axiomatised and thus forced into cogency that one realises that they are not about measurements and that they do not deal with observers and their mental states. One realises then that the first theory is concerned with autonomously existing physical systems that can be connected by electromagnetic waves. And one realises that quantum mechanics is about microsystems eventually acted on by macro-systems, which are further physical things rather than observers. The physical quantities one computes in both theories must therefore be given strictly objective meanings. And if this is the case with the two theories that have allegedly brought the subject or observer back into the picture of the universe, we can trust that the whole of physics is nowadays as concerned with the external world as it was in Galilei's time.

In a well-built theory every possible referent (body, field, or quantum mechanical system) is mentioned at the start: it occurs in the list of basic or undefined concepts. The adjunction of a *deus ex machina*, such as the observer, to the physical systems concerned becomes then logically impossible in such a context. It is only by arbitrarily introducing extraneous elements at the level of theorems, i.e. by smuggling in concepts that did not occur in the axioms, that nonphysical (subjectivist) interpretations pop up. In short: whatever concept is to be used in a theory has to be introduced as a primitive or else defined in terms of primitives. Since neither the observer nor the (nonexistent) general purpose apparatus are either primitive or defined concepts in special relativity and quantum mechanics, they do not belong by right to these theories. If theories of measuring instruments and of measurement processes are to be built, they must be built as applications of all the theories actually involved in the measurements concerned. More on this in Chapters 4 and 10.

In sum: the analysis of theories is best conducted in an axiomatic context: open context analysis is bound to be defective. And this holds also for the philosophical analysis of theories and for the analysis of the philosophical theses about theories: the best way to evaluate any philosophical claims about a theory (for instance that it confirms a certain epistemological tenet) is to behold the theory as a whole and stripped of every appendage which is not needed either for making computations or for applying the theory to real situations. This does not mean that the foundations of physics can be philosophically neutral and must be absolutely prior to the philosophy of physics: it does mean that there must be a mutual adjustment of the two. Foundations without philosophy is illusory, and philosophy without foundations is superficial and often irrelevant.

5. MEETING PLACE AND BATTLEGROUND

The foundations of physics, and particularly the axiomatic reconstruction of physical theories, are a suitable field for the cooperation of theoretical and mathematical physicists, applied mathematicians, logicians, and philosophers of physics. Indeed, all these skills are called for and no single individual is likely to master them all, so that cooperation is mandatory. It is to be hoped that this opportunity will not be missed.

What holds for cooperation holds for debate: when it comes to dissenting on philosophical issues and points of methodology, the foundations of physics offer an adequate battleground. I can think of no better way of settling the question whether a given physical theory is committed to a certain philosophical thesis, than axiomatising the theory and checking whether that thesis is in fact contained in the theory either explicitly or as a presupposition. Surely this method is superior to either rhetoric or authority. What a pity it is used so seldom. What a pity most of us prefer to debate fundamental questions in the same unscientific and unphilosophic fashion we debate ideological and political matters.

So much for the salient traits of the philosophical and foundational approach to physical theory. Let us now characterise the structural, semantical and methodological aspects of physical theories.

NOTE

* Some paragraphs are reproduced from Bunge (1970b) with the permission of the editor and publisher.

CHAPTER 3

PHYSICAL THEORY: OVERVIEW

Because the main concern of both the foundations and the philosophy of physics is the analysis and reconstruction of physical theories, it will be fitting to start by characterising a physical theory in general terms. At first sight this is a light task: in physics a theory is really nothing but a mathematical formalism endowed with a physical interpretation and capable of coexisting with other physical theories as well as of being checked by experiment. This looks fair and sounds simple but it is actually complex. Consider the following questions posed or evaded by the previous definition:

– What if any are the assumptions and theories a given physical theory takes for granted: just mathematical theories or others as well?

– Is the formalism of a physical theory uniquely determined by the key formulas we want to systematise or are there alternatives and if so are they equivalent in every respect?

– What is meant by 'physical interpretation': a visualisable model, a mechanical analogue, the reference to laboratory operations, the reference to external objects, or what?

– What is meant by the coexistence of theories: just their logical compatibility or also a partial overlap and therefore a mutual assistance and check?

– How is the phrase 'experimental testability' to be construed: as concerning every formula in a theory and the whole range of every formula, as the possibility of being contradicted by empirical data, as the possibility of numerous confirmation, or what?

These are only a few typical representatives of a populous set of questions sparked by the very concept of a physical theory. Every one of them can be made as deep and tough as desired and many of them cannot be answered except by a whole paper or even an entire book. Which exemplifies the general rule: *What is obvious to the practitioner of a science may be problematic to its philosopher.* Since we cannot take up every possible foundational and philosophical problem concerning physical theories in general, we shall select a few topics for consideration in this book. And, in order to forestall endless misunderstandings, let us begin by fixing the terminology and by clearing the ground for future chapters.

1. SOME KEY TERMS

In contemporary philosophical, mathematical and scientific parlance a *theory* is not just a stray opinion but a hypothetico-deductive system, i.e. a set of formulas generated by a bunch of initial assumptions with the help of logic and mathematics. By virtue of the generality of some of those initial assumptions, as well as by the transformation possibilities afforded by logic and mathematics, every theory is an infinite set of formulas. Just think of all the possible situations covered by a universal law, be it as modest as Archimedes' law of the lever; and think of all the derivatives of any given function occurring in the theory. If only for this reason, i.e. because every theory is infinitely rich, a conclusive proof of it is out of the question. The best we can do is to confirm a theory in a large number of cases or to refute it at some critical points. Even so, refutation can be dodged and thus confirmation courted, not only by refusing to admit adverse but suspect evidence but also by adjusting some of the components of a theory, notably the values assigned certain parameters. We shall touch on this in Chapter 10.

Some of the initial assumptions of a physical theory are called *hypotheses* (in the epistemological not in the logical sense). Hypotheses, whether particular or general, go beyond a mere description of observable situations: they are conjectures about matters of fact, whether or not these situations are partially observable. Thus it is a hypothesis of mechanics that there are bodies, another that mass is conserved, a third that the body stress is representable by a real and bounded tensor field. Some of the hypotheses explicit or implicit in a physical theory are of a purely mathematical nature in the sense that they stipulate the mathematical characteristics of the concepts concerned – e.g. the symmetry of a tensor. Others have a more or less direct physical meaning, in the sense that they concern properties of real systems or of systems conjectured to exist in reality.

The most important among the physical hypotheses of a theory are of course the *law* statements. A law statement purports to concern objective patterns or modes of existence and change of physical systems. It does not convey information about particular situations and it does not tell us what the world looks like for some observer: a law statement is supposed to be universal and observer-free. Equations of motion, field equations, constitutive equations and equations of state qualify as laws to the extent to which they belong to reasonably confirmed theories. Further physical hypotheses are *subsidiary hypotheses*, such as initial conditions, boundary conditions, and constraints.

Every physical hypothesis is supposed to be mathematisable. But mathematical form alone won't tell us anything about the physical meaning of the formula. Thus the formula "$E_n = -k/n^2$" could mean anything. A formula with no fixed physical meaning may be said to be semantically indeterminate, i.e. indeterminate as to its meaning. It will become semantically determinate upon adjunction of extra assumptions, usually tacit, concerning some of the symbols involved in it. Thus in our previous example 'E_n' could stand for the energy of a hydrogen-like atom in the nth level. In a different context the same typographical mark would "acquire" (i.e. would be assigned) an altogether different meaning. Such additional assumptions as sketch the physical meaning of symbols can be called *semantic assumptions*.

Data, i.e. statements obtained by observation or experiment, constitute a further kind of initial assumption. They are initial in the sense that they must be assumed in order to yield some logical consequences or theorems. But, of course, data are not supposed to be made up, i.e. they are not a priori. Nor do they have to be obtained by experience alone: rather on the contrary, the data that can be entered in a physical theory have to be couched in the terms of the theory and they must be obtained with the help of instruments designed and read with the assistance of further theories. In short, data are not given but sought and, if relevant to a physical theory, they are soaked with theory rather than being a direct expression of the observer's sensations or feelings.

A fourth kind of premise occurring in a theory is the *definition*, e.g. of the electric field energy density:

$$\rho_E =_{df} (1/8\pi) E^2 .$$

Formally considered, a definition is just a linguistic convention, i.e. a rule for handling the symbols concerned: it tells us nothing about nature.

This conventional nature of definitions does not render them arbitrary: what can be defined in a theory and what cannot is something to be decided when reconstructing the theory in an axiomatic fashion. And the choice of defining (undefined, basic or primitive) concepts must be guided by definite criteria such as generality and fertility.

Note the difference between definitions and hypotheses, in particular law statements: while the former inter-relate concepts the latter relate statements to reality. Hence while definitions can be subjected to conceptual criticism alone, law statements are supposed to live up to experimental tests as well. However, this elementary distinction is often forgotten. For example, Mach's celebrated construal of classical mechanics can be traced back, at least in part, to his failure to distinguish hypotheses (like Newton's law of motion) from definitions (see Bunge, 1966).

So far, then, we have the following kinds of formula to be found in any physical theory ready to be applied to solving some problems:

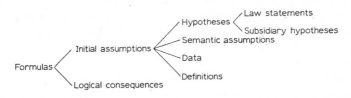

Whatever its status, a physical formula is a *statement* expressed by a *sentence* belonging to some language. (Sentences are linguistic objects. Statements are conceptual objects. One and the same statement can often be expressed by quite different sentences.) A statement is treated, for purposes of logical processing, as if it were true or false. As regards its correspondence to fact, we may not know what the truth value of the statement can be; at other times we rate it low but not minimal, and occasionally we give it at a high mark but not maximal. Statements (or propositions) obey a calculus of their own: the propositional calculus and, more generally, the predicate calculus. This calculus systematises the rules of deductive inference, such as:

$$Pa \vdash (\exists x) \, Px$$

which can be read thus: That the individual a has the property P entails that there exists at least one individual exemplifying that property. Notice that P is a dummy: it stands for any property whatsoever, whether physical or nonphysical. The predicate calculus is a branch of logic, the science presupposed by every other rational discipline and which no experiment can refute. The reason for this aloofness is that logic is not concerned with the world but with statements and their transformations quite apart from their content. Nevertheless, it is fashionable to claim that, just as general relativity necessitated a change in geometry, so quantum physics obeys its own logic. This is mistaken. All of the quantum theories utilise ordinary mathematics, which has ordinary logic built into it. One of the sources of the mistake is to take literally a formal analogy between propositions and projection operators (von Neumann, 1932). No matter what algebra a family of operators may obey, the statements in the algebraic theory obey ordinary logic: an operator is a concept not a statement.

Just as sentences are decomposable into words, so statements can be analysed into concepts. The concepts occurring in physics are either formal or factual. The *formal* concepts are all those borrowed from logic and mathematics. The *factual* concepts of physics are peculiar to it. They are factual in the sense that they concern real or conjectured matters of

---

Theory: Maxwell's electromagnetic theory for free space.

Formal concepts (tacit or overt) *involved in physical hypothesis below:*
 differentiable manifold, vector and pseudovector valued functions on the former, partial derivative, vector product.

Basic (undefinded) physical concepts involved in hypothesis below:
 Physical space, time, E, B, c
 Defined physical concepts: $\nabla \times E$, $\partial B/\partial t$.
 Operational definitions: none.

Hypothesis: Faraday's law of electromagnetic induction in its differential version:

$$\nabla \times E = -(1/c)\,\partial B/\partial t.$$

Subsidiary assumption: E and B drop off with distance at least as fast as $1/r$.

Semantic assumption: E represents the electric strength, B the magnetic induction and c the speed of light in vacuum.

Datum: none.

fact. A factual concept need not be empirical, i.e. it need not concern an observational or experimental situation. Moreover, to qualify as a physical concept, a concept must not revolve around an observer: it must concern a possible physical system, situation, or event. More on this in Chapter 4.

The preceding block illustrates some of the concepts we have discussed in this section.

2. THE MATHEMATICAL COMPONENT

The role of mathematics in modern science is dual: concept formation and computation. No concept of instant velocity without the derivative concept, no law of motion without differential or operator equations. Mathematical concepts are not just handy auxiliaries: they are the very core of physical ideas. And the simplest prediction of a future state of a system or of the probability of an event would be impossible without the deductive power inherent in the formalism of a theory. This deductive power is so formidable, and effective calculation so time-consuming, that we often tend to equate theoretical physics with computation, forgetting the role of mathematics in the very formation of physical concepts, formulas, and theories.

Computation techniques, though indispensable, are not physical theories: they are not even self-sufficient mathematical formalisms. A calculation method (e.g. for diagonalising matrices) is part of a mathematical theory which may (but need not) be in turn part of the formalism of a physical theory. By itself mathematical theories are neutral with respect to any hypotheses about the actual world. Consider the theory of canonical transformations, once regarded as the nucleus of quantum mechanics. Whether in its classical or in its quantal forms, it is not an independent physical theory portraying some aspect of the world: it is a mathematical method for solving equations of motion (Hamilton's, Schrödinger's, etc.) and for relating the solutions obtained in different representations. The whole point of the theory is to simplify the statement of a problem, hence its solution, while preserving the equations of motion and certain invariants. So much so, that the theory can be applied irrespective of the physical content of the equations.

Likewise perturbation theories can be applied in a number of fields as long as one has got a definite equation to which they can be applied. That is, they carry no physical meaning: they are useful mathematical auxiliaries, means for attaining an end, which is the approximate solution of some equation that may have some physical meaning. One or two terms in a perturbation expansion can be assigned a physical content each: the infinitely many terms in the whole series could not possibly be interpreted. The meaning neutrality of perturbation methods can also be seen by analysing the concept of order of an effect. Q: What does this expression mean: does it tell us anything about nature? A: Nothing about nature, only something about our computation technique. Thus a fourth order effect is one accounted for by a theoretical model involving a perturbation expansion up to the 4th degree, i.e. neglecting all higher powers (even if the series diverges). The same effect might be accounted for by different theories ascribing different orders – or none, for one such theory might give an exact solution. The same holds of course, for any series expansion and every decomposition of a vector into components: while the function as a whole may be physically meaningful, the mode of decomposition is purely mathematical and can be changed almost at will.

Physical content, if any, must be sought in some of the concepts and statements of a theory, not in a particular *representation* of properties and laws. For example, one and the same trajectory in ordinary space may be written out in any kind of coordinates. Every coordinate transformation generates a new representation without changing physical contents. So much so that the only valid restrictions on representation changes brought about by coordinate transformations are (i) the transformed variables should have the same meaning as the old ones (e.g. the Lorentz transform of a position coordinate should be a position coordinate not a time coordinate); (ii) the transformed variables should obey the same law statements as the old ones. What holds for coordinate systems holds also for units. While the representation of a physical property by a function involves a choice of units, these are conventional, and consequently shifts in units carry no physical meaning.

That some components of a physical theory are devoid of physical content is less surprising than the possibility of assigning a physical meaning to others. Indeed, we have become accustomed to the idea that mathematics is devoid of physical content. We first learned that a continuous function need not be a time-dependent magnitude; later we learned that geometry is noncommittal unless it is superimposed semantic assumptions. Some people have still to learn that arithmetic and probability theory are equally neutral and therefore must be adjoined semantic assumptions if they are to become applicable, but by and large we have come to understand that mathematics is an autonomous discipline even though many mathematical ideas have been motivated by scientific research. Yet for all its purity mathematics is applied in physics – or, as our predecessors used to say, "mathematics is applicable to reality". Q: How is this possible? A: While every symbol occurring in a physical theory has some mathematical meaning, some mathematical symbols are assigned, *in addition*, a physical interpretation. Thus 'dX/dt' can be interpreted not only as the total derivative of a certain function X, but *also* as an instantaneous rate of change of some physical property represented by X, such as a position coordinate, a concentration, an energy, or what have you. A physical content thus rides on a mathematically meaningful sign and in this way both horse (or donkey) and knight roam across the physical arena. (All this can of course be said in nonmetaphorical terms. To say it is a task of the semantics of science: see Bunge, 1972a, b.)

A physical concept differs from its underlying mathematical concept in two respects: (i) every physical concept concerns some physical system(s) and (ii) every physical concept enters at least one physical law. By contrast, purely mathematical concepts have no extramathematical referents and they obey no extramathematical laws. Take for instance the relation "heavier than or as heavy as", or H. From a formal point of view H is nothing but an ordering relation $\geqslant$ on some set B of unspecified elements: i.e. $H \subset B \times B$ and $H \in$ Set of ordering relations. H becomes a physical concept when (i) B is interpreted as the set of bodies and (ii) H is assumed to be connected in B, i.e. to hold among any two bodies.

The case of the weight function is even more instructive because there are infinitely many equivalent ways of representing the physical property of weight (or any other physical property), namely one per system of units. The weight of a body $b \in B$ in a gravitational field $g \in G$, relative to a (physical) reference frame $k \in K$ and reckoned on the unit $u \in U_W$, is a certain nonnegative number w; i.e. $W(b, g, k, u) = w$. Weight in general is the function W itself rather than any of its values. And this function maps the set $B \times G \times K \times U_W$ of all quadruples $\langle b, g, k, u \rangle$ with $b \in B$, $g \in G$, $k \in K$ and $u \in U_W$ (the set of weight units) into the set R^+ of nonnegative real numbers:

$$W : B \times G \times K \times U_W \rightarrow R^+ .$$

Furthermore (and here comes the law) W is such that $W(b, g, k, u) = m\ddot{X}$, where m is the mass and $\ddot{X}$ the acceleration of the body b. (Our tacit restriction to the nonrelativistic particle model of a body is of course immaterial in this context.) Every other magnitude has a similar structure: it is some function from a cartesian product with at least two factors, one a set of physical systems of some kind, the other a set of units.

Very often one of the sets of physical systems occuring in the domain of a magnitude is the set of reference frames of some kind, e.g. the frames relative to which Newton's laws of motion hold. Such frames are sometimes called "observers" in compliance with an observer-centered philosophy, namely operationalism. But obviously observers are neither as ubiquitous nor as dumb as reference frames and in any case their study does not belong to physics. In sum, physical meanings are infused into a formalism via basic physical quantities representing properties of physical systems and obeying physical laws.

The preceding analysis disqualifies numerology as a serious approach to physical theory. Numerology can be defined as the juggling with dimensionless constants (pure numbers) with a view to producing significant relations. Because numerology is concerned with dimensionless constants alone, it is hard to assign it a physical content. And because the number game can be played on a computer without feeding it any law statements, numerology cannot lead, except by accident, to physical laws. Moreover, it is trivial, as shown by the following

THEOREM. Given n nonnegative numbers $a_1, a_2, ..., a_n$, there exist infinetely many n-tuples of nonzero real numbers (positive or negative) b_1, $b_2, ..., b_n$ such that

$$a_1^{b_1} \cdot a_2^{b_2} ... a_{n-1}^{b_{n-1}} = a_n^{b_n} .$$

(Proof: first take logarithms and look into $n=2$. Then apply mathematical induction.) Once a given n-tuple of exponents has been found, it is easy to approximate every one of them by a simple fraction. So a "striking" relation will have been produced. The procedure can then be repeated with a different choice of exponents, and so on *ad infinitum*. Our success in finding such a numerical combination depends on our ability and resources. No knowledge of any laws of physics is required. This does not invalidate numerology as an occasional and weak heuristic tool: number games can occasionally lead to insights and even spark a theory proper. But the point is that numerology is no theory and it involves no physical laws. This had to be stressed because every time undigested data accumulate (as is the case with particle physics and cosmology these days) there is a tendency to try to juggle with them rather than risking deep hypotheses that might account for such data.

So much for the role of mathematics in physics. Let us now turn to the other end of the spectrum, namely data.

3. THE EMPIRICAL COMPONENT

Two extremes have usually been avoided in theoretical physics: one is the *a priori* theory that wants and needs no data, the other is the theory that accepts all possible data, even conflicting ones. Every genuine physical theory, however false, makes room for some data: mutually compatible data of a given sort, i.e. concerning physical systems of certain kinds and in definite states. And every physical theory can produce new possible data, i.e. predictions – or retrodictions – in unlimited number if enriched with suitable particular premises. (See Section 4.) A theory devoid of predictive power cannot be used and therefore cannot be subjected to empirical tests.

The predictive ability of every physical theory worth its name is so remarkable that it has nourished the instrumentalist view that scientific theories, far from being pictures of the world, are nothing but data churning devices. This popular view is utterly false: if a theory did not concern reality and if it included no law statements it could not make predictions or its predictions would not be borne out. In other words, the data combinations effected by scientific theories must not be either arbitrary (like those gotten with a randomiser) or magical. If this restriction is observed, then a scientific theory may indeed be regarded, for purely practical purposes (i.e. leaving explanation aside), as a data factory.

A physical theory must accept some actual data as inputs and must be able to generate from them another set of possible data (the output) in such a way that both input and output match the assumptions of the theory – laws, constraints, etc. This concept of matching involves *rele-*

vance: thus boundary conditions are relevant only to field-like theories such as hydrodynamics and quantum mechanics. But matching is more than relevance: it is also logical *compatibility*. Thus the (imaginary) datum "The speed of propagation of the *F*-field is infinite", even if true, would be inconsistent with the (imaginary) *F*-field theory and therefore could not be entered in it to calculate anything. A particular form of compatibility is *instantiation*: thus an initial entropy is an instance of an entropy value and thus compatible with any theory containing the entropy concept as long as the values of the other physical quantities linked with entropy are not fixed independently. As with the data input so with the data output: it will match the theory if the calculations are correct. This process cannot help generate significant potential data because of the laws included in the theory. In short, the logical schema is:

$$\{Theory,\ Data\} \vdash Predictions.$$

This schema holds irrespective of the kind of physical theory: it may be phenomenological like electric network theory or it may involve some mechanism like a theory of conductivity in solid state physics. It may be stochastic like quantum mechanics or nonstochastic ("deterministic") like general relativity. Note also that, while the set of data in the above schema is finite, the set of predictions is potentially infinite: what we get at the end of a calculation in theoretical physics is not just a bunch of numbers but a set of related functions. And the ranges of these functions have usually at least the cardinality of the natural numbers.

No magic in this prolificacy nor induction either: although the data fed into the theory are in finite number, at least some of the hypotheses are universal: they are taken to hold for all possible objects of a kind, all possible values of an "independent" variable, and so on. This universality is so much taken for granted that we usually neglect to write the appropriate quantifiers in front of the equations concerned. No such neglect is admissible in an axiomatic formulation. For example, when writing out in full Faraday's law of electromagnetic induction we should prefix it the following phrase: "At any point in the space time manifold, for every electromagnetic field and every charged body, there exists a reference frame such that" – and here Maxwell's first triplet follows.

The preceding account of the theory-data interdependence implies a rejection of the view that theories are just data summaries, at most slight extrapolations beyond data. If theories were nothing but data summaries they would hardly anticipate data, let alone in infinite number, much less data different in kind from those fed into them, as when fields are calculated from charges and currents. Besides, if a theory were just a data concentrate it could not possibly conflict with any data. And we would be unable to understand (explain) anything with the help of theories: for a heap of data, no matter how high, is something to be explained not an explanation. The dataist view ignores not only the nature of theories but also their role in eliciting data. In physics a datum is usually a sophisticated item that cannot even be stated unless some theories are accepted.

Take for instance a track in a bubble chamber or in a nuclear emulsion. In order to interpret it as a track produced by a particle we must hypothesise: (a) that there has been such a particle, (b) that the particle was electrically charged, for this is what our theories require for a particle to leave a track, (c) that the particle can interact with the material and (d) that the hypothetical particle satisfies at least the law of energy and momentum conservation – for this alone will enable us to read some numbers out of the measured quantities (length and density of the tracks). If this conservation theorem were not assumed, it would be impossible to detect neutral particles and estimate their mass values: the hypothesis is, indeed, the theoretical basis of the missing mass method. Any failure of the observed tracks to conform to the energy and momentum conservation theorem could be attributed either to the failure of the latter or to the presence of one or more neutral particles that carried away a fraction of the input momentum. Experimental physicists trust theory at least on this point: they assume that the hypothesis holds and so are able to discover (or rather discovent) a number of neutral particles.

The role of theories in experiment is then no less important than the role of empirical data in activating theories and testing them. Thus the astronomer needs optics to design and correct his telescopes; similarly the particle physicist needs theories explaining the functioning of his detectors: otherwise he might be counting his own heart beats; and anyone using a galvanometer or even a humble scale is trusting the theory of his instrument. Any such theory concerning an experimental set-up or a component of it may be called an *instrumental* theory, while the theory that is being activated or tested is a *substantive* theory. We shall dwell on this in Chapter 10.

The referent of an instrumental theory is thus an artifact, such as a nuclear emulsion, rather than a natural object, such as a cosmic ray. No artifacts are referred to specifically by substantive theories even if they are general enough to cover some aspects of many an instrument. However, some extremely general theories such as the relativistic theory of gravitation and quantum mechanics are often expounded in terms of instruments and measurements, such as clock readings and diffraction through a slit system. Such references are phony, because general theories are uncommitted to special equipments. In particular, relativistic theories do not talk about clocks but about time: if they did concern clocks then (a) they would have to contain particular assumptions about real clocks of some kind (pendulum clocks, atomic clocks, laser clocks, etc.) and (b) no special theories of clocks would be needed: it would be possible to use relativistic theories for computing, say, the periodic momentum transfer to the pendulum, the friction as a function of the velocity, and other characteristics of any wrist watch. For better or for worse, a clock theory is a very special application of a general theory, usually mechanics. The same holds for the phony measurement devices occurring in the operationist formulations of the quantum theories: such instruments could not be conceived and operated without quantum mechanics and several additional theories.

The interdependence of theory and experiment refutes the popular view according to which physics (and science in general) is like a fruit with a hard core surrounded by a soft pulp: the core would be a set of data and the pulp the theories built around the former. Both the core and the pulp would be in continual growth (the core leading the pulp) but, while the former would grow cumulatively, theories would be nibbled away by fresh experiments. This view can be refuted in two ways: by counterexamples and by showing that it does not match the actual method of research. As to the former, suffice it to mention the following: (a) ever since the "violation" (refutation) of the joint conservation of charge and parity was announced, an oscillating sequence of measurements have resulted in alternatively positive and negative experimental evidence; (b) in solid state physics, where sample purity is essential (for traces of an impurity may result in macroproperties), disagreement among equally competent experimenters is frequent: solid state data are no less fluid than others.

Concerning the methodological untenability of the fruit analogy, two points should suffice. The first is that experimentalists handle real materials, which are rarely pure, and they manipulate them in the midst of an active and polluted environment made of air and an assortment of fields. Since such conditions cannot always be controlled or even detected, different workers are bound to obtain different results under "the same" conditions for, as a matter of fact, no given condition can be replicated exactly. All the experimentalist can do is (a) to remove more and more sources of discrepancy (e.g. to improve insulation) and (b) to record more and more exactly what the actual conditions are and (c) to correct for them with the help of theories. Even so data, if accurate, are bound to be somewhat discrepant: exact consensus smacks of chance coincidence. The second point of method concerns the relevance of theory to experiment. Interesting empirical information is produced in the light of some theory (however embryonic) and with the help of some instrumental theories.

In short, data can be as controversial as theories. No lasting harm can result from the softness of both data and theories if they keep checking one another. This permanent possibility of bilateral correction is more typical of science than either erratic trial and error, or cumulative growth, or total revolution.

Let us sum up our discussion of the theory-data interrelations:

(i) *Data can stimulate the creation of theories.* They will do it if they are anomalous (at variance with some theory) or if, having been obtained with reliable instrumental theories, none of the available substantive theories makes room for them.

(ii) *Data can activate theories.* By entering data into a theory, the latter can be made to yield specific explanations and predictions.

(iii) *Data can test theories.* By confronting theoretical predictions with data, the truth value of the former can be assessed. However, data can hardly pass any judgement by themselves: the verdict of further, adjoining theories must also be heard.

(iv) *Theories can guide the search for data.* First, by predicting previously unknown effects. Second, by helping design experimental set-ups.

(v) *Blind data are as useless as stray ones.* Unless well confirmed theories have participated in producing a datum, it should not be trusted. And unless the datum is consistent with at least some well corroborated hypotheses it is a curio that could be explained by some systematic error in the design of the experiment or the reading.

(vi) *Theories have no observational content.* Empirical information (e.g. initial temperatures) must be fed into a theory from the outside if any further potential information (prediction) is to be deduced. Consequently (a) theories cannot be inferred from data and (b) physical theories cannot be interpreted in empirical terms (e.g. in terms of length and time measurements) but (c) they must be interpreted in objective physical terms, i.e. with reference to observer-free physical systems.

4. GENERAL THEORY AND MODEL

Continuum mechanics is an extremely general theory: it depicts bodies of all kinds. It is so general that it can solve no particular problem unless adjoined special assumptions concerning the system concerned. On the other hand particle mechanics is a special theory – so special that it can solve problems of a few kinds. And the classical theory of the harmonic oscillator is an even more special one – a theoretical model of a free vibrator. Similarly the general theory of quantised fields is so general that with it alone not a single cross section can be calculated. On the other hand quantum electrodynamics is a more specific theory and the Compton effect theory an even more special one, namely a theoretical model of the elastic scattering of photons by electrons. In both cases the special theory, and in particular the theoretical model of a system, is obtained by adjoining subsidiary assumptions to a general framework – for example by specialising the hamiltonian or by adding constitutive equations (material laws). In summary,

{General theory, Special assumptions} ⊢ *Special theory.*

In young fields no general frameworks are available: one has at most a theoretical model, i.e. a special theory covering a narrow species rather than a wide genus of physical systems. And, whether or not a general theory is available, if one wishes to deal with a specific kind of things, e.g. liquids in turbulent motion or atomic nuclei shelled by protons, one must construct a model of them, i.e. an idealisation or sketch of the real thing, one that seizes on some of its characteristic traits. In other words, a theoretical model of a system includes a model of it, i.e. a schematic

TABLE I

System	Model object	Theoretical model	General theory
Moon	Spherical solid spinning about its axis, rotating around fixed point, etc.	Lunar theory	Classical mechanics and gravitation theory
Moonlight	Plane polarized electromagnetic wave	Maxwell equations for the vacuum	Classical electromagnetism
Piece of glass	Random linear chain of beads	Statistical mechanics of random chains	Statistical mechanics
Crystal	Lattice plus electron cloud	Bloch's theory	Quantum mechanics

representation of the real or conjectured system. This model is sometimes called a *model object*. Table I illustrates some of the ideas we have just discussed.

Let us take a look at the first example. When special assumptions and data concerning a particular body are adjoined to classical mechanics and classical gravitation theory, a special theory about that body is produced. Thus we have lunar theories, Mars theories, Venus theories, and so on. The lowest level statements of these theories are expressions for the co-ordinates (spherical geocentric) of the body concerned. These functions solve the equations of motion and they consist of Fourier series. In order to get numerical values one has to assign time a definite value and add up the series. The summation process is usually approximate: one keeps only a finite number of terms in the expansion. Any discrepancy between the special theory and observation outcomes can be imputed either to errors of observation or to some of the ingredients of the theoretical model. Usually the discrepancies are attributed to neglected terms in the series expansion. This is the case of the celebrated "gravitational inconsistencies" (a strange misnomer) in current lunar theory, discovered as recently as in 1968. It would be foolish to look beyond for the discrepancies – e.g. in special or general relativistic effects. One trusts the general theoretical framework – but then only because, if conjoined with special assumptions, it rarely leads to such discrepancies with data. However, in principle all the ingredients are suspect: the general theory, the special assumptions, the model object, the computation, and even the data. Only the underlying mathematical formalism is above suspicion – if, indeed, it is consistent.

No specific computations, hence no confrontation with data, without some model object, or sketch of the physical system concerned. A *model object*, joined to a set of law statements and other premises, yields a *theoretical model* of the real thing. Calling the latter R and a model of it M, we can write: $M \triangleq R$, read '*M* represents *R*'. Any such representation is partial: it does not (and should not) cover every single trait of the represented object. Conversely, some traits of the model M may correspond to nothing in its referent R, i.e. they may be redundant. The partial nature of the model-thing correspondence is well illustrated by the two simplest (poorest) model objects: the mass point and the black box. The mass point or particle is not a thing but a model of a body. It may be construed as a list (ordered *n*-tuple) with the following items: a point in ordinary space, mass, and velocity. (All its other properties are derivable from these.) And the black box concept may be regarded as a system-environment pair together with three functions: input, transducer, and output.

In either of these two cases the shapeless and structureless model, be it a point or a black box, is loaded with the responsibility of representing a natural system with a shape and a structure which are either ignored or actually irrelevant to the purpose at hand. The details of the represented object are thus lost or deliberately obliterated. Take a closer look at this partial correspondence in the case of the particle-body relation:

MASS POINT	BODY
Point position	Region of space
Point velocity	Velocity field
Mass	Mass distribution
Force on mass point	Body force
_____	Contact force
_____	Stress
_____	Electric current distribution
_____	(E,B) field
_____	(D,H) field
_____	Temperature distribution
_____	Entropy density
_____	Etc.

If a continuous body model M' is substituted for the mass point model M as a picture of the real body, we get another model or representation of the same object and consequently an alternative theoretical model of it. Any of the continuous body models M' (with or without electrodynamic and thermodynamic properties) is richer than M: there is a function from the list M to any of the lists M' but not conversely. In general: of two model objects M and M' of a physical system R, M' is *more complex than M* if and only if there is a correspondence f from M *into* M'. And two models M and M' of a given concrete object R are equally complex iff there is a correspondence f from M *onto* M'. The more complex models are not necessarily the truer ones but they have better chances of being so.

No model object is the exclusive property of a given theory. For example the mass point may be assumed to satisfy any number of equations of motion: it may thus be shared by a number of theoretical models. In general: any given model object can, within bounds, be embedded into a variety of alternative theories. Indeed, since a model object is just a list of traits, these traits can be characterised and mutually related in an unlimited number of ways to produce as many theoretical models. Conversely, any general theory can be adjoined alternative model objects as long as the latter are built with concepts occurring in the general framework. (This proviso is important yet often forgotten in connection with quantum mechanics: many a conceptual difficulty in this theory stems from the stubborn attempt to graft onto it classical model objects such as those of particle and wave.)

The foregoing has important methodological consequences. A first consequence is this: the empirical refutation of a given theoretical model does not constitute a refutation of the underlying general theory if there is any. Example 1: the slight inaccuracies in the relativistic theory of the gravitational field of the sun should be blamed on the Schwarzschild solution, which rests on the point mass model of the sun. Example 2: the failure of the available theories of nuclear forces to give satisfactory explanations of the stability, structure and transformations of atomic nuclei does not refute quantum mechanics: it may be blamed on the particular models (e.g. hamiltonians) that have been assumed so far.

A second methodological consequence of the distinction between general theory, theoretical model and model object is this: strictly speaking, general theories are untestable. Indeed, by themselves they can solve no particular problem, hence they cannot generate any specific predictions.

Only theoretical models can be confronted with data. For example, general continuum mechanics is untestable without further ado. On the other hand particle mechanics, a very special subtheory of it (a theoretical model) is testable. (But being special it cannot generate the general theory, notwithstanding the valiant textbook efforts to build bodies out of points.) In short, only special theories (theoretical models) are testable, and they are so by virtue of containing definite model objects (Bunge, 1969a).

Even so, we must recall that no testable theory is *fully* testable. Firstly because it is impossible to check every one of its infinitely many statements. (Recall Section 1.) Secondly because even a low level theorem, say a solution to a field equation, cannot be checked for every value of the "independent" variables, among which the object variables, i.e. the ones representing the systems concerned, must be counted. Thirdly because any given set of data is consistent with an unlimited number of alternative high level formulas. Even a given set of law statements can be subsumed under widely different axioms: thus any given set of equations of motion can be derived from any number of alternative lagrangians. Fourthly because every powerful theory has a number of statements that keep dangling far too high above the ground of experience – such as the formulas for the position and velocity of an electron in an atom. In sum, while theories can be supported or undermined by spot checking they cannot be proved. Even their disproof is a complex (though not impossible) business due to the number of more or less uncertain components at stake

This ultimate uncertainty concerning the worth (truth value) of scientific theories has inspired an antitheoretical bias that is often expressed in the attempt to cleanse theories from their transempirical or nonobservational ingredients. Such a programme is not viable and this by definition of 'scientific theory': a scientific theory is a hypothetico-deductive system: that is, a system based on hypotheses, i.e. statements that transcend observations in the sense that they concern whole classes of facts not just those that happen to be observable. Besides, observability, or rather measurability, depends on theory: without theories the most interesting and accurate data would not be forthcoming. (More on this in Chapter 10.) The progress of science does not consist in a progressive elimination of unobservables but in their proliferation and taming. A scrutable unobservable, one that is somehow linked to observable effects, is at least as valuable as a directly manipulable variable and surely far more valuable than any number of observations untutored by theory.

To conclude let us list the main types of problem the theoretical physicist is likely to face:

(i) Given a collection of data, find a set of data-covering formulas. Feel free to invent nonobservational concepts as long as they are scrutable.

(ii) Given a set of data-covering formulas, glue them together into a theory. Feel free to invent far-reaching hypotheses as long as they are in the main susceptible to empirical confrontation.

(iii) Given a collection of special theories (theoretical models), find a general theory. Feel free to give up some special hypotheses and generalise others.

(iv) Given a general theory, adjoin it special assumptions to obtain a theoretical model. Feel constrained by the actual problem at hand.

(v) Given a theoretical model, obtain a set of predictions. Feel bound by the actual data.

(vi) Given a set of predictions, watch their performance and draw inferences about the worth of the premises. Feel free to modify the latter as needed, as well as to discard uncertain data.

We have come full circle: experience sparks theoretical problems, some of whose solutions take us back to experience. Every stage in this cycle is just that: a stage that has no value separated from the other stages. We need this reminder at a time when the physical profession is split into instrument makers, experimentalists, theoretical physicists with an eye on experiments, theoretical physicists with an eye on mathematics, mathematical physicists, and foundations workers. A merit of philosophy is that it reminds us of the whole beneath that division of labour.

This completes our overview of physical theories. The subsequent chapters will focus on a number of special foundational and philosophical problems concerning physical theories. The first of them will be the question: What are physical theories about? This will involve us in the debates over realism, subjectivism, and conventionalism, which have marked the philosophy of physics over the last hundred years.

CHAPTER 4

THE REFERENTS OF A PHYSICAL THEORY*

It is generally recognised that theoretical physics has been in an impasse for at least two decades. In particular, no breakthrough has been made in the domain of "particle" physics: there are no general theories with a predictive power in this field. A number of formidable technical difficulties stand in the way, but there are also certain philosophical obstacles that could easily be swept aside. The chief among them is the current confusion and uncertainty concerning the referents of fundamental physical theories, i.e. the kind of thing they are about (Bunge, 1971a) For, if such theories are about language, as it is occasionally claimed, then obviously we must turn to linguistics for guidance through the maze. If they concern propositions, then we should ask logic to supply the answers to the pressing questions of particle physics. On the other hand if every theory about microsystems is about an object-apparatus-observer block that cannot be further analysed (Bohr's "essential wholeness of a proper quantum phenomenon" [Bohr, 1958a, pp. 72 and passim]), then obviously there is no finer analysis to be expected. But if, on the other hand, physics must introduce the observer's mind as a separate and determinant factor into the picture of the world (Wigner, 1962; Heitler, 1963; Houtappel *et al.*, 1965), then there is hope for progress as long as physics joins forces (or rather failings) with psychology. And if, finally, physics is about things that are presumed to be out there, then its task remains the traditional one of getting to know more and better about them rather than either proclaiming final victory or turning inward, to the study of the self. In any case identifying the referents of a physical theory is not only of philosophical importance: it concerns the very strategy of scientific research.

1. THE INTERPRETATION PROBLEM

1.1. *The Referent*

That which a construct (concept, proposition or theory) is about, or stands for, or refers to, or rather is intended to refer to, is called the (*intended or hypothetical*) *referent* of the construct. The referent of a construct may be a single object or any number of objects; it may be perceptible or imperceptible; it may be real, or imaginary, and so forth. In any case the referent of a construct is a collection of items and is therefore also called the (*intended or hypothetical*) *reference class* of the construct. For example, the reference class of the concept "cold" is the set of bodies and the one of "The earth spins" is {Earth}, while the referents of Dirac's new positive energy relativistic wave equation have so far not been identified.

Some reference classes are *homogeneous*, i.e. composed of elements of a single kind – e.g., deuterium atoms. Other reference classes are *inhomogeneous*, i.e. composed of elements of distinct kinds, such as protons and synchrotons, or atoms and external fields. A reference class composed of entities of kinds A and B may be construed as the union $A \cup B$ of the corresponding sets. A construct will be said to *refer* to a class A just in case the set A is included in the reference class of the construct.

Our problem is to find out the nature of the reference class of a physical theory. In particular, we want to ascertain whether the reference class of a physical theory is made up, if only in part, of cognitive subjects – e.g., observers – or of their mental states. For there is little doubt that some of the expressions occurring in physical writings do refer legitimately, at least partially, to cognitive subjects. Any such expressions will be called *pragmatic expressions*, in contrast to *physical object expressions*, which are free from any reference to cognitive subjects. For example, "The value of the property P for the physical object x equals y" is a physical object sentence (or rather sentence schema), while "Observer z found the value y for the property P of the physical object x" is a pragmatic sentence schema. Clearly, while in the first case reference is made only to a physical system, in the latter case there are two referents : a system and an observer. Or, if preferred, while in the first case it is a question of the *value* of a physical quantity, in the other case it is a question of an *observed value* or empirical estimate of the same quantity, i.e., of its value for a given observer. This difference may look tiny but it is both scientifically and philosophically significant, as will be shown shortly. Suffice now to show that, whereas the preceding physical object sentence has the form:

$P(x)=y$, the corresponding pragmatic sentence may be analysed as: $P'(x, z)=y$ or, even better, as : $P'(x, z, t, o)=y$, where t stands for the measurement technique and o for the sequence of operations employed by the observer z in implementing that technique when measuring P on x. We have chosen the new symbol P' to designate the measured property because it stands for something patently different from P: indeed, while the function P is defined on a set X of physical objects, the function P' is defined in the set $X \times Z \times T \times O$ of quadruples physical object-observer-measurement technique-sequence of acts. We shall return to this point in Section 2.

It is hardly disputable, then, that the language of physics does contain pragmatic sentences. (And also pragmatic metasentences like "Nobody knows whether the quark hypothesis is true.") What is being controverted is the thesis that *all* the nonmathematical sentences occurring in every physical *theory* are pragmatic sentences in the sense that at least one of the components of the reference class of a physical theory is a set of human subjects, such as qualified observers. In other words, what is still a matter of opinion among physicists is the semantical problem of identifying the referent of a physical theory either as a physical system, or a subject, or a subject-object synthesis or, finally, as a subject-object pair. In short, what is still controversial is the interpretation of physical symbols and, in particular, the interpretation of the formulas of theoretical physics: are they physical object sentences, or else mental object sentences, or perhaps physico-mental sentences or, finally, partly physical and partly mental object sentences? Before rushing to pick the right answer we ought to know what the interpretation possibilities are.

1.2. *Interpretation*: *Strict and Adventitious*

However arbitrary theological hermeneutics may be, the interpretation of scientific formulas should not be a matter of choice. To begin with, the interpretation assigned to the basic or undefined symbols of a scientific theory should not render the latter inconsistent and should turn it true – or, more realistically, maximally true. (For example, it would be wrong to interpret the square of the wave function as a mass density, for this would be inconsistent with the normalisation condition.) Second, if a symbol is defined or derived in terms of previously introduced signs, then its meaning should "flow" from the latter rather than being concocted *ad hoc*.

(For example, it is wrong to interpret the time derivative of the average of a position coordinate as an average velocity unless the variable can be interpreted as representing a physical position, which is far from obvious in relativistic quantum mechanics.) Third, a strict interpretation of a complex expression should be compatible with the structure of the latter; in particular, if it is claimed that a certain complex symbol concerns a thing of a given kind, then at least one of the constituents of the symbol must be capable of denoting that particular thing. (For example, in order for the wave function to concern both a microsystem and an apparatus, it must actually depend on variables of the two of them, which is the exception rather than the rule.)

The previous conditions seem obvious yet they are often ignored. The first condition is just pointless in relation with most theory formulations, for it applies only to axiom systems: indeed, it is only in an axiomatic context that the basic/defined dichotomy makes sense. The second condition is violated whenever a defined quantity (or a derived formula) is interpreted in terms that are alien to the defining terms (or to the premises, as the case may be). This condition is violated, for example, when the entropy of a physical system, computed on the basis of data and assumptions concerning the system itself, is interpreted as a subject's amount of information concerning the system, even though no premises concerning the subject and his fund of knowledge are supplied. As to the third condition, its violation is exemplified by the following typical case. If someone claims that a formula such as "$y=f(x)$" concerns the f-ness (whatever this property may be) of an object x of a certain kind X, such as observed by an observer z of some kind Z, then he is introducing a ghostly variable, namely z. This variable (and the whole set Z) is phony because footless: in order for something to count as a genuine referent it must hold the reference relation to some sign, and in the above case no such symbol corresponding to the alleged observer z occurs in the given expression. (As will be seen in Section 3.2, the standard quantum theory of measurement contains such phony variables.)

An interpretation of a (nonformal or descriptive) variable shall be called *strict* if it assigns the variable just one object. If every nonformal symbol in an expression is assigned a strict interpretation, the expression will be said to be interpreted in both a strict and *complete* way. If at least one of the symbols but not all of them are interpreted in a strict way, then the interpretation will be called strict and *partial*. Any interpretation, whether partial or complete, that is not strict will be called *adventitious*. For example, the interpretation of the symbol "$v(x, y)$" as the velocity of a system x is strict and partial; its interpretation as the velocity of a system x relative to a reference frame y is both strict and complete; and its interpretation as the velocity of a system x relative to a frame y as measured by an observer z with the technique t, is adventitious as far as the variables z and t are concerned.

Clearly, a strict and complete interpretation is preferable to either an incomplete or a redundant one: we should neither fail to read some of the components of the meaning of a symbol nor read too much in it. However, not all strict interpretations produce true formulas and not all adventitious interpretations produce false ones. That a strict interpretation can lead to falsity is shown by interpreting, say, the formulas of thermodynamics in terms of subjective probability. That an adventitious interpretation can be true, even trivially so, is equally clear: thus in the example of a function f relating two variables, if f is the right one and the experimenter does his job properly, he will obtain values of f close to the calculated ones. But the experimenter may bungle, in which case the adventitious interpretation becomes false. Also, thus stressing the role of the experimenter may create the impression that the object owes it f-ness to the former, in which case the adventitious interpretation becomes misleading. No such risks are run by strict interpretations. Therefore we must cast a cold eye on adventitious interpretations.

1.3. *Pragmatic Interpretations*

Since the birth of operationism there has been a strong tendency to construe all linguistic expressions in pragmatic terms. This happens not only in relation to formulas susceptible to empirical tests but also in relation with mathematical formulas. This is common practice in the classroom, where it has some didactic virtues, and is the mark of mathematical intuitionism – the mathematical partner of physical operationism. Yet all such pragmatic interpretations of mathematical symbols are adventitious, for a point of mathematics is to abstract from users and circumstances in order to achieve both universality and freedom from commitment to fact. Take, for instance, the asterisk symbol used for complex conjugation. A strict interpretation of the expression '$z*$', where z designates a complex number, in this: 'z^*' means the real part of z minus i times the imaginary part of z. (This rule of designation might be replaced by a definition.) By contrast, a pragmatic interpretation of the same symbol is this: "Anyone presented with the symbol 'z^*' is supposed to reverse the sign of the imaginary part of z." A second pragmatic interpretation of 'z^*' is in the form of a rule or prescription: "To compute z^* from z, reverse the sign of the imaginary part of z," A third pragmatic reading of the same symbol would be an instruction capable of being fed into a computer so as to enable it to handle the symbol. Every pragmatic interpretation of a logical or mathematical symbol may be construed as an instruction for handling (e.g., computing) the symbol in an effective way.

Given a mathematical symbol, it can be assigned any number of pragmatic interpretations, according to the user, the circumstances, and the goals (e.g., in connection with different kinds of computers). This plurality of pragmatic interpretations is possible because they are adventitious whenever they concern mathematical symbols: that is, they are not subjected to the internal mathematical laws the symbols themselves satisfy. (Indeed, pure mathematics tells us nothing about users, circumstances, or goals.) It is therefore to be expected that only a subset of the conceivable pragmatic interpretations of a mathematical symbol will be valid. In any case, we need a validity criterion. We shall presently propose one applying to both mathematical and factual symbols.

We stipulate that a pragmatic interpretation of a sign be *valid* just in case there exists a theory containing that sign and such that it provides a ground or reason for the procedure indicated by the pragmatic interpretation. Thus arithmetic provides a ground for (it justifies) the instructions given children for operating an abacus, as well as the instructions fed to a computer in order to find, say, a given power of an integer. A pragmatic interpretation will be called *invalid* just in case it is not valid. Thus interpreting a chemical formula in pragmatic terms involving incantations rather than, say, mixing, stirring, and heating, would be open to the charge of invalidity, for there is no theory countenancing a relation between chemical structure and incantations. Likewise, the interpretation of entropy as a measure of our ignorance is invalid, for it involves the erroneous identification of statistical mechanics with epistemology. (Note that "valid" does not amount to "right." Thus, an invalid or unjustified pragmatic interpretation may eventually prove to be right, for a theory may be built that justifies it. Conversely, a valid pragmatic interpretation may turn out to be wrong, for the theory supporting it may have to be given up.)

Whether the formulas of theoretical physics may be assigned valid pragmatic interpretations, remains to be seen (see Section 2). What is hardly disputable is that pragmatic interpretations are at home in *experimental* physics, where the reference to observers and circumstances of observation is legitimate and often explicit. Here we find two kinds of pragmatic interpretation: strict and adventitious. Let us start with the former. An expression such as "$f(x, y) = z$" can in principle be interpreted in this way: "The f-ness of x, as observed (or measured) by y, equals z." Since the formula makes room for the pragmatic referent y, the preceding interpretation is both strict and partly pragmatic. But, of course, any such interpretation must be only *partly* pragmatic if it is to count as a sentence in the language of physics, for this science happens to be concerned with physical systems. Secondly, the variable y must designate a possible observer, not a mythical one like the observer at infinity (or even worse, the continuous stream of observers) imagined by some field theorists. Thirdly, y must be a variable in the intuitive sense: that is, a change in the value of y must make some difference to the value of f. In short, subject to certain restrictions, *some* physical formulas can be given a *strict* interpretation which is *partly pragmatic*.

But the most common pragmatic interpretations found in the physical literature of our century are adventitious rather than strict: that is, they do not assign a meaning to every variable in a complex symbol but take the latter in block and match it with a pragmatic item from the outside. Thus given a sentence s belonging to a physical language, the following pragmatic interpretations of s are frequently met with in the literature: (*a*) "s summarises the measurements performed by a qualified observer"; (*b*) "Perform the operations necessary to test s"; and (*c*) "Act (analyse, measure, build, destroy...) in conformity with s." The reference to the physical object has been obliterated: everything points now to an active subject. Consequently the ideal of scientific objectivity seems to have been discarded.

While a strict pragmatic interpretation may contribute to determining the meaning of a symbol, an adventitious interpretation, whether pragmatic or not, fails to perform this function: it just prescribes or suggests, in a more or less precise fashion, a way of action. It does not tell us what the symbols stand for but what can be done with them – which is all technicians and Wittgensteinian philosophers care for. Secondly, for a sign *s* to be handled in an effective way and in conformity with one of its pragmatic interpretations, *s* need not make full sense to its prospective user even though it must make some sense to the person ultimately responsible for such use. Thus computations and even measurements can be executed with the help of computers that are given only pragmatic interpretations. But the programmer must be aware of the semantic interpretation of the symbols he handles, for otherwise he will be unable to write out any program and decode the machine's output. Thus if a sentence *s* expresses some property *P* of a physical system, any pragmatic interpretation of *s* for the use of, say, an automated measurement arrangement, requires not only an adequate semantic interpretation of *s* (i.e., one pointing to physical systems), but also its linking to a number of further sentences capable of expressing a way in which *P* can effectively be measured on a concrete physical system. (These additional sentences usually belong to theories other than the theory in which *s* inheres.) In other words, the design and execution of empirical operations, whether automated or not, involves the assignment of *pragmatic interpretations based on semantic interpretations*. In short, adventitious interpretations, even when legitimate, cannot replace semantic interpretations.

1.4. *Four Theses Concerning the Referent of Physical Theory*

Before asking what the referent of a theory may be we must ask whether it does have a referent at all. There are two possible answers to this previous question: "Do physical theories have a referent?" One is in the affirmative, the other in the negative. The latter is, indeed, the conventionalist or instrumentalist view according to which physical theories are not about anything but are just data summarising and processing tools, i.e., instruments enabling us to can information and grind out predictions. This answer is unsatisfactory for at least two reasons. First, it fails to tell us what kind of data physical theories are supposed to handle and what sort of predictions they are supposed to yield in contradistinction to, say, sociological theories. Second and consequently, a consistent conventionalist would not know how to go about testing a physical theory, for every such test presupposes knowing what the theory is supposed to be concerned with : indeed, if a theory intends to refer to say, fluids, then it will be fluids rather than, say, atomic nuclei or wars which will have to be looked into. We shall therefore discard the conventionalist thesis.

A nonconventionalist should then have an answer to the question: "What are physical theories about?" Since the referent of a factual theory may be a physical object, or a subject, or some combination of the two, there are four possible and mutually exclusive answers to the question of the identity of the referent. These are:

(1) The *realist* thesis: a physical theory is about physical systems, i.e., is concerned with entities and events that undoubtedly have an autonomous existence (*naive realism*) or else are assumed (perhaps wrongly in some cases) to have an autonomous existence (*critical realism*). In short, the physical interpretation of every nonformal formula in theoretical physics must be both *strict* (as opposed to adventitious) and *objective* (as opposed to subjective): every theoretical statement in physics is thus a *physical object statement*.

(2) The *subjectivist* thesis: a physical theory is about the sensations (*sensism*) or else about the ideas (*subjective idealism*) of some subject engaged in cognitive acts – in any case it is about mental states. In short, the physical interpretation of every formula in theoretical physics must be both *strict and subjective*: every theoretical statement in physics is thus a *mental object statement*.

(3) The *strict Copenhagen thesis*: a physical theory, or at any rate quantum theory, is about unanalysable subject-object blocks. No absolute (subject-free, objective) distinction can be drawn between the two components of any such block: the boundary between them can be shifted at will. In short, the physical interpretation of every nonformal formula in theoretical physics, or at least in quantum theory, must be both *adventitious* (as opposed to strict) and *physico-mental* (as distinct from either physical or mental), for the observer and his conditions of observation must be read in every such formula even though the corresponding variables may be missing: every theoretical statement in physics is thus a *physico-mental* statement.

(4) The *dualist* thesis: a physical theory is about both physical objects and human actors: it concerns the transactions of humans with their environment (*pragmatism*) or the ways humans handle systems when intent on knowing them (*operationism*). In short, the physical interpre-

tation of every formula in theoretical physics, whether *strict or adventitious*, must be *partly objective, partly pragmatic*: every statement in theoretical physics is thus a *partly physical and partly mental object statement*.

The realist thesis is the one that prevailed during the classical period of physics, and wich was defended by Boltzmann, Planck, the later Einstein, and de Broglie. The subjectivist thesis was frequently defended by Mach (whose "elements" or atoms were sensations) and occasionally also by Eddington and Schrödinger. The Copenhagen thesis was advanced by N. Bohr and defended by his faithful followers – the qualifier being in order, for most of those who profess their loyalty to the Copenhagen school actually waver between (3) and (4) above. And the dualist thesis has been expounded in various ways by Peirce, Mach, Dingler, Dewey, Eddington, Bridgman, Dingle, and Bohr, as well as by hundreds of writers on relativity (who identify reference frames with observers) and quanta (who take apparatus for observers). The fourth thesis is, indeed, the nucleus of the official philosophy of physics (recall Chapter 1), even though it has never been vindicated by a careful analysis of theoretical expressions.

The first three theses are *monistic* in the sense that each of them asserts that the reference class of a physical theory is metaphysically homogeneous (physical, mental, or physico-mental respectively) and moreover irreducible to entities of a different kind. The fourth thesis is *dualistic* in the sense that it postulates two mutually irreducible substances. Mathematically speaking, the reference class of a theory interpreted in a monistic spirit is homogeneous in the sense that all of the individuals or members of that set are assumed to be of the same broad kind: either physical, psychical, or psychophysical (see Section 3.1). On the other hand, the reference class of a theory cast in a dualistic spirit will be the logical sum of two or more sets, at least one of which represents a kind of physical object while at least one other such set represents observers. (See, however, Section 3.3 for the impossibility of carrying out the dualist program.) Thus, e.g., the domain of an absolute probability function occurring in a physical theory will be interpreted in the following ways by the various semantical schools encountered in the philosophy of physics: the set of physical events of a kind (realism), the set of mental events of a kind (subjectivism), the set of irrational (unanalysable, hence incomprehensible) phenomena of a kind (Copenhagen), and the set of pairs physical event-observer (dualism).

Every monistic thesis expresses a commitment to the hypothesis that there are entities of a certain kind: physical objects, minds, or psychophysical complexes, as the case may be. The dualist, on the other hand, while claiming that physics is about both things and actors, will refuse to acknowledge the *independent* existence of physical objects, and will therefore come close to the Copenhagen doctrine. He will advance the methodological thesis that a statement concerning a thing in itself is untestable. And, basing himself on the verification doctrine of meaning that was fashionable four decades ago, he will conclude that such a statement is meaningless. The pragmatist, then, is neither a realist nor a subjectivist: he is an agnostic just like Kant. And, like the Copenhagen philosopher, the pragmatist maintains that a theoretical statement makes no sense unless it is accompanied by a description of the conditions of its empirical test. But, unlike the Copenhagen philosopher, the dualist distinguishes the subject from the object and he does not deride the attempt to analyse the subject-object interaction even though he may not care to perform it himself. Furthermore, the dualist is not prepared to acknowledge the existence of a third kind of stuff composed, in arbitrary varying proportions, of object and subject.

To discover which of the four preceding philosophical theses about the content of a physical theory is right, quotations from the Old Masters will not help, not only because arguments from authority are worthless but also because, as we saw before, every one of the four above theses enjoys the support of at least one great name. Worse: one and the same author may endorse two mutually incompatible theses in the same writing, without seeming to realise their difference. Thus Mach and Dewey wavered betweeen subjectivism and dualism; and Bohr, who started out as a realist, became a subjectivist (Bohr, 1934), oscillated later between dualism (as often represented by Heisenberg) and the strict Copenhagen thesis (Bohr, 1958a), and seems to have finally reverted to realism (Bohr, 1958b). Nor will general philosophical discussions be of much help, for the object of our inquiry is a special kind of human knowledge. Rather, we must take the bull by the horns and analyse physical theories and their components (concepts and statements). We shall proceed to sketch such an analysis with particular reference to quantum theory, for, while the problem antedates this theory, it has become more acute with its birth.

2. IDENTIFYING THE REFERENT

2.1. *The Theoretical-Experimental Dichotomy*

Although the four theses expounded in the last section concern the reference of a physical theory, all but the first, or realist thesis, are meant to cover the whole of physics, both theoretical and experimental. Indeed, for the consistent subjectivist every self-contained physical formula is a mental object sentence; for the Copenhagen philosopher every such expression concerns an indissoluble mind-body compositum; and for the dualist every such sign is about both objects and subjects. On the other hand the realist claims that, while an empirical statement (e.g. an experimental datum) concerns both a physical object and an observer (or a team of observers), a theoretical statement fails to point to any subject whatever, the concern of theoretical physics being to account for the world such as it is, independently of its being perceived or manipulated. In short, of all four only the realist makes a semantic difference between theoretical and experimental physics.

Moreover, the realist will probably point out that this distinction enables him to mark off the meaning of a formula from its test – two concepts that are hardly distinguishable for both the Copenhagen and the dualist philosophers. And he may add that this distinction between theoretical and experimental physics makes it possible to understand why theoreticians work only with their heads while experimentalist must, in addition, fumble around with pieces of equipment. Finally, the realist may also say that the very same distinction is necessary to understand why theories are not self-testing (as they should if subjectivism were true) and why any empirical test consists in contrasting and assaying theoretical predictions, on the one hand, and experimental results on the other. In any case, whether or not we end up by adopting realism, we should give it a chance to defend itself, accepting to make the theoretical-experimental distinction even if we intend to end up by denying that there is any.

Now, theories are certain sets of statements (infinite sets closed under deduction), and every physical statement contains at least one physical concept, for otherwise it would not qualify as a physical statement. Therefore our semantic analysis of physical theories must begin with physical concepts. It may as well end with them, for this level of analysis is necessary and sufficient to disclose the referent of a physical theory: a theory is indeed about all and only those items referred to by the concepts with which the theory is built. A systematic investigation of this kind being out of the question in the present context, we shall restrict our attention to a few typical examples.

2.2. *The Referent of a Physical Quantity*

The so-called physical quantities (or rather magnitudes) are said to "pertain" to a physical system of some kind or to be "associated" with such an object. This association of symbol with thing becomes obvious when it is necessary to name the components of a complex system, e.g., by assigning them numerals. Thus the P-ness of the nth component of a system may be designated by P_n.

These vague phrases may be elucidated by introducing two basic semantic concepts: those of referent and representation. Indeed, what is meant by saying that P "pertains to" or "is associated with" a physical system of a certain kind, is this: the concept (function) P represents a physical property (call it $\mathscr{P}$) of an arbitrary system σ of a certain kind, call it Σ. Hence Σ is the (intended) *reference class* of P. (In a moment we shall generalise this to the case of multiple referents.) The explicit mention of the referent is a reminder that, unlike functions in pure mathematics, those in theoretical physics may concern actual physical systems. And the explicit mention of the representation of a property $\mathscr{P}$ by a concept P calls our attention to the possibility that one and the same property (e.g., the electric charge) may be represented by alternative concepts in different theories. In short: P *refers* to Σ and *represents* $\mathscr{P}$. All this will presently be clarified and exemplified.

Let P be a function representing the property $\mathscr{P}$, or $P \triangleq \mathscr{P}$ for short. Now a function is not well-defined unless both its domain and its range are given. In the simplest case, of an invariant quantitative property, the function in question will be "defined" on a set of elements interpreted as so many physical systems, and into some set of numbers .The domain of this function will be either a set of individual systems (as in the case of the charge) or a set of pairs (as in the case of an interaction) or, in general, of n-tuples of physical systems. These sets are, of course, the reference classes of the concept P.

Example 1. In classical electrodynamics, the electric charge is represented by a function Q from the set Σ of material systems to the set R^+ of non-negative reals, i.e., $Q: \Sigma \to R^+$. Actually a further "independent variable" occurs in Q, namely the scale-cum-unit system s, which is usually specified by the context of the formula in which Q occurs. Hence, a correct analysis of the classical electric charge function is:

(1) $\qquad Q: \Sigma \times S \to R^+$

where S is the set of all conceivable scale-cum-unit systems. For example, for Σ=electrons and s=electrostatic units on a uniform metric scale (for an explication of a physical scale, see Bunge, 1967c) one has the (usually unspoken) law statement:

(2) $\qquad$ For every σ in $\Sigma: Q(\sigma, \text{esu}) = e \doteq 4.802 \times 10^{-10}$.

In any case, the reference class of Q is the set Σ of bodies.

Example 2. In all physical theories, the position (or else the position density) of a point of a physical system, whether it be a body or a field, a classical or a quantum-mechanical system, is represented by a vector valued function X of the elementary system, the reference frame, and time. In short, presupposing again a uniform metric scale,

(3) $\qquad X: \Sigma \times F \times T \to R^3$

where F is the set of physical frames and T the set of instants. Since T is not a thing, the reference class of X is just $\Sigma \cup F$. This is what is meant by the definite description "the position of σ relative to the frame f."

Example 3. The effective cross section of a particle of kind A for elastic scattering by a particle of kind B with wave momentum k relative to a frame f (e.g., the center of mass system), is a function σ (notice the change in notation) from the set of quadruples $\langle a, b, f, k \rangle$, where $a \in A$, $b \in B$, $f \in F$, and $k \in K$, to the positive real line. In short,

(4) $\qquad \sigma: A \times B \times F \times K \to R^+$

where K is the momentum range. For example, for A=protons, B=neutrons, and f=center of mass frame, we have, to a first approximation, the well-known quantum-mechanical formula

(5) $\qquad \sigma_{PN}^{cm}(k) = 4\pi/k^2$.

Actually the CGS system has been presupposed. In any case, the reference class of σ is $A \bigcup B \bigcup F$.

Our semantic analysis so far favors a monistic interpretation: indeed, the reference class of a physical magnitude seems to be either a set of systems or the union of sets of systems. The observer is nowhere in sight. Moreover our analysis refutes the Copenhagen doctrine, insofar as no traces of the subject-object sealed unit have been detected. Only two views are upheld by our analysis: realism and subjectivism (see Section 2.2). The choice between the two cannot be made on the basis of our analysis alone, for the subjectivist will find no difficulty in identifying every system we call "physical" as a mental object.

Only an analysis of the sum-total of scientific activities tilts the balance in favor of realism (Bunge, 1967c). Let the following reasons suffice here. (a) Every investigator starts out by admitting his ignorance of something he presumes, if only provisionally, to exist by itself and waiting, as it were, to be discovered. (b) Every properly formulated theory starts by assuming that the reference class it is concerned with is nonempty, for otherwise the theory would be vacuously true. But this is no less than a (critical) commitment to the hypothesis that the theory has real referents. (c) No matter what subjectivist fancies the theoretician may indulge in when writing popular articles, the experimentalist is bound to take a realist attitude towards his experimental arrangements, the objects of his inquiries, and even his colleagues.

But, as we mentioned before, the observer concept, while absent from theoretical physics, does enter experimental physics. Thus, instead of the observer-free formula (2), we find in experimental physics statements like this: "The value (in esu) of the electron charge e, as measured by the experimental group g with the technique t and the instrument complex i, equals $(4.802 \pm 0.001) \times 10^{-10}$." In short, instead of (2) we now have

(2')   For every σ in Σ examined by $g : Q'(\sigma, \text{esu}, g, t, i)$
$$= (4.802 \pm 0.001) \times 10^{-10},$$

where Q' designates the function whose values are measured values of the electric charge. Note that Q' differs from Q (the theoretical concept) not only numerically but structurally as well: it is a *different concept* altogether. In fact, while Q is a function on the set $\Sigma \times S$, Q' is "defined" on the set $\Sigma \times S \times G \times T \times I$, where G is the set of experimental groups, T the set of charge measurements techniques, and I the set of charge measurement equipments. Actually a sixth "independent variable" is involved in the experimental concept of electric charge, namely the sequence o of operations by which any given technique t is implemented by a group g with the help of the instrumental complex i. That o is not a phony or vacuous variable (in the sense of Section 1.2), is shown by the fact that changing its value usually does make a difference in the numerical result. Calling O the set of all such sequences of operations, we have finally, in contrast to (1),

$$(1') \qquad Q' : \Sigma \times S \times G \times T \times I \times O \to \mathscr{P}(R^+),$$

where $\mathscr{P}(R^+)$ is the set of all intervals of R^+.

We now feel justified in drawing a general conclusion from the preceding analysis: *While the theoretical formulas are observer-free, the experimental formulas are observer-dependent.* More precisely: whereas any strict interpretation of a theoretical formula is *objective* (i.e., is cast in terms of physical concepts alone), experimental physics calls for a *pragmatic reinterpretation* of the same formula. Such a pragmatic interpretation, though adventitious (yet possibly valid) if bearing on a theoretical formula, is a strict interpretation in an experimental context, i.e., it is validated by substituting an experimental formula (such as $(1')$) for its theoretical partner (e.g., (1)).

Both the Copenhagen and the dualist interpretations of physical theories arise from a confusion between theoretical and experimental concepts, even though the latter rest on the former and are more complex than their theoretical bases. (In particular, a theoretical function may be regarded as the restriction of the corresponding measured value function to a certain set. Thus the Q in (1) is the restriction of the Q' in $(1')$ to $\Sigma \times S$.) This confusion may not be deplored by the Copenhagen philosopher, for whom everything is incurably irrational at bottom, but it defeats the very aim of the dualist, which is to avoid Platonizing. Indeed, if every theoretical formula is assigned a pragmatic interpretation, then it becomes impossible to contrast theory to experiment in order to test the former and design the latter. Moreover, since pragmatic interpretations are mostly adventitious, arbitrariness is apt to set in : everyone will feel entitled to read any formula as he pleases, independently of the structure of the formula: semantics will have no syntactic support. The realist and the subjectivist interpretations are free from these shortcomings. If we drop subjectivism for the reasons indicated a while ago, realism remains as the sole survivor. We should adopt then a *strict and objective* interpretation of every theoretical formula. Let us see how this approach fares in quantum theory, often alleged to be the grave of realism (and causality).

2.3. *State Vector*

It is generally agreed that the state vector or wave function ψ is a probability amplitude (i.e., that the square of its modulus is a probability density). Moreover, this, which is Max Born's "statistical" (actually probabilistic or stochastic) interpretation of ψ, can be proved from a certain set of postulates, whence it is far from being *ad hoc*, hence avoidable if the standard formalism of quantum mechanics is kept (Bunge, 1967a, pp. 252 and 262). On the other hand there is no consensus about what ψ is a probability amplitude of. Sometimes ψ is regarded as concerning an individual system, others as referring to an actual or a potential statistical ensemble of similar systems, others as measuring our information or our degree of certainty concerning the state of an individual microsystem-apparatus complex, or finally as summarising a run of measurements on a set of identically prepared microsystems. (For a critical survey of a number of interpretations of the state vector, see Bunge, 1956 and 1959b.) In any case it is customary to assign ψ some such interpretation without caring whether it matches the structure of ψ and without making sure that the interpretation contributes to the consistency and the truth of the theory.

Yet it is possible to avoid the arbitrariness inherent in the adventitious interpretations of ψ and spot its genuine referents. The key is of course the hamiltonian operator H since, according to the central law of quantum mechanics (Schrödinger's equation or its operator equivalent), H is what "drives" ψ in the course of time. Now, H is not given from above but is what we want it to represent: the state of a carbon atom, a DNA molecule, or what not – and this in any hamiltonian theory, whether classical or quantal. We must then start by stating our claims or hypotheses concerning what it is that H, hence ψ, is going to be about. Some such claims will prove to be (approximately) true, others false: such is the life of theories.

Let us consider the mathematically simplest (but semantically most problematic) of cases: one-"particle" (or rather one-quanton) quantum mechanics. If the formalism turns out to be true under a certain interpretation of the basic (primitive) symbols, then the whole thing, formalism-cum-semantics, will be judged true of such individual systems even though its empirical test will call for the intervention of sentient beings manipulating (directly or by proxy) large collections of microsystems. In this theory H, hence also ψ, depends on time and on two sets of dynamical variables: those concerning the microsystem of interest (e.g., a silver atom) and those concerning the environment of the system (e.g., a magnetic field). If no such macrosystem is assumed to act on the given microsystem, i.e., if the latter is assumed to be free, then no macrovariables will occur in the hamiltonian (nor, consequently, in the state vector), no matter how much *ad hoc* talk of observers and measuring devices may be indulged in. Indeed, it is utterly arbitrary, hence a matter of blind belief, to claim that even though a given hamiltonian fails to contain macrovariables, actually it concerns an observing mind, or a mind-body compositum, or even a microsystem-apparatus complex. Any such interpretation, by failing to match the syntax of H and ψ, is adventitious: it involves vacuous or idle variables: it has a ghostly quality. Both the formalism of (elementary nonrelativistic) quantum mechanics and the set of its applications (e.g., to molecules) warrant only this analysis of every state vector:

$$(6) \qquad \psi : \Sigma \times \bar{\Sigma} \times E^3 \times T \to C$$

where Σ is the set of microsystems, $\bar{\Sigma}$ the set of macrosystems, E^3 the ordinary (Euclidean) space, T the range of the time function, and C the complex plane. Consequently the reference class of this theory is $\Sigma \cup \Sigma'$, i.e., the union of microsystems and their environments. Any other interpretation is without foundation: it has no leg to stand on except the *dicta* of some famous scientist and their philosophical apologists. (For a rich collection of authoritative quotations in support of the Copenhagen interpretation and a return to Bohr's ideas, see Feyerabend, 1968. For criticisms see Hooker, 1972.)

2.4. *Probability*

Of all the adventitious interpretations of the state vector, the subjectivist or nearly-subjectivist one is the most resilient, so that it will be worth our while to consider it in some detail. A popular argument for the thesis that ψ must be subjective or at least partly so, is the following: "The state vector has only a probability meaning [*true*]. Now, probabilities concern only mental states: a probability value can only measure the strength of our belief or the accuracy of our information [*false*]. Hence the state vector concerns our minds rather than autonomous physical systems [*false*]." This argument is valid but its conclusion is false because its second premise is wrong: indeed, a task of stochastic theories in physics is to compute *physical* probabilities (e.g., transition probabilities and scattering cross sections) and statistical properties (averages, mean scatters, etc.) of physical systems, not of mental events. In any case, it is surprising that the same scientists who have often or even consistently adopted a subjectivist interpretation of probability, such as Bohr, Born, Heisenberg, and von Neumann, have at the same time believed that they had overcome classical determinism, an essential ingredient of which is the thesis that probability is but a name for ignorance. It would seem that, ever since statistical physics and statistical biology were born, we have acknowledged randomness as an objective mode of becoming, formerly only of aggregates, now of individual entities as well. In any case, the subjective interpretation of probability has no place in physics and it presupposes classical determinism.

Whether or not probabilities are to be interpreted as physical properties on a par with lengths and densities, is not a matter of opinion but one of mathematical and semantical analysis. Only an examination of the independent variable(s) of a probability function will tell us whether the function can be assigned a (strict) interpretation as a physical property, or as a state of mind, or as "pertaining" (referring) to some thing-mind complex. But although such an analysis will indicate the interpretation possibilities it will not suffice to find out whether any of them is an admissible interpretation. The latter will be the case only if the probability calculus, or rather the whole formalism including that calculus, becomes factually true under the given interpretation. And, again, this is not a matter of taste, or philosophical school, or arbitrary decision, but rather a matter to be settled by analysis and experiment.

Take, for instance, the expression '$Pr(x)=r$', where 'Pr' stands for the probability function and r for a member of the real number interval [0,1]. If x stands for a physical object, such as a state or a change of state, then $Pr(x)$ will be a property of that object, and any reference to observers, their operations, or their states of mind, will be supernumerary. Only if x symbolises a psychical event will '$Pr(x)$' stand for something mental. There is no place for two referents, e.g., a physical object and a psychical one, where there is a single independent variable. Absolute (unconditional) probabilities are then impregnable to strict pragmatic interpretation in terms of both physical objects and actors: in order to rope in a subject we need a further variable. This possibility is afforded by conditional probabilities.

The expression '$Pr(x \mid y) = r$', read as 'the probability of x given y equals r', could be interpreted in either an objectivist, or a subjectivist, or finally a dualist way (namely as concerning a thing-subject couple). For example, if the context or, event better, the explicit interpretation rules and assumptions, indicate that x stands for a physical object (e.g., a state or an event) and y for an observer, then "$Pr(x \mid y)$" might be read as the probability that the physical event x happens provided the observer y is present, or given that the mental event y has occurred, or in some other dualist fashion. But, as remarked earlier, any such interpretation will be legitimate just in case (a) the theory contains both independent variables and specifies them, and (b) the theorems of conditional probability are satisfied under this interpretation, i.e., they are satisfactorily confirmed by observation. There is actually a third condition to be met as well, namely the one of relevance: one can always add an observer variable, but unless this variable does make a difference and its properties are specified by the theory, it will be a vacuous or phony variable. So much for the strict interpretation of probability.

A pragmatic interpretation is always possible, even for unconditional (or absolute) probabilities, and it is often necessary, but it is never strict, i.e., it does not "flow" from the formulas but must be superimposed on them in a way that overflows physical theory. What I mean is this. Surely some probability values must be checked by someone, either theoretically or empirically, or both ways, so that statements of the following kind ensue: "The probability value r for event x was checked by observer y with the means z." But this statement does not belong to the theory: it does not qualify as a strict interpretation of the formula "$Pr(x) = r$." Something similar holds for every physical property, not just for probability. Thus the physical statement: "The distance between the end-points x and y of the body z, as measured by the observer u with means v, equals

$r \pm \varepsilon$." In short, given a theoretical statement with a strict physical interpretation, it can be attached any number of adventitious pragmatic
interpretations. But none of the resulting pragmatic statements belongs
to the theory, just as none of the parasites of a tree is a part of the tree.
The popular operationist contention that only pragmatic statements are
meaningful, so that meaning assignments require a reference to empirical
operations, rests on a confusion between meaning and verification, a
confusion that has long since been cleared up by philosophers.

2.5. *Interpretation and Estimation of Probabilities*

It is widely held that the frequency interpretation of probability, i.e., the
interpretation of probabilty values as relative frequencies, is what we
want in science. But this is not quite so. Indeed, when reading probabilities in terms of relative frequencies we do not perform a strict interpretation but rather a valuation or (statistical) *estimation*. That is, we do not
assert that probabilities *mean* frequencies but that they can (sometimes)
be *measured* by frequencies. In this regard a probability does not differ
from any other physical quantity: it is a construct whose numerical
value is to be contrasted to a measured value. Moreover, as there is no
unique measurement technique for any given physical magnitude, so
there is no single way of estimating probabilities from statistical data:
sometimes one counts frequencies, at other times one measures entropies,
at some other times one measures spectral line intensities, sometimes one
measures scattering cross sections, and so forth. The very theory in
which the probability concept is embedded may (but usually does not)
suggest ways of estimating probabilities. In most cases additional
theories are needed to estimate probabilities from empirical data. But
this is not peculiar to probability: it holds for other properties as well.
(See Chapter 10.)

There are five additional reasons for rejecting not only the frequency
theories of probability (like von Mises' and Reichenbach's), which are
mathematically untenable anyhow, but also the frequency *interpretation*
of probability. First, what one seems to mean by '$Pr(x) = r$' in physics is
something like the strength (measured by the number r) of the tendency
or propensity for x to occur, quite apart from the number of times it is
(actually or potentially) observed to happen. The latter count will serve
the purpose of testing a probability formula rather than the one of assign-

ing it a meaning. Second, while probabilities can be properties of individuals (e.g., events), frequencies are collective properties, i.e., properties of statistical ensembles. Third, the formulas of probability theory are not satisfied exactly by frequencies, not even in the long run, which is always a finite run. (Remember that frequencies do not approach probabilities. Only the *probability* of any preassigned departure of a frequency from the corresponding probability decreases with increasing sample size. But this theorem holds only for a special kind of random process, namely a sequence of Bernoulli trials. Furthermore, the second order probability that the theorem is concerned with is not itself reducible to a frequency.)

Fourth, probability and frequency *are not the same functions*, for whereas the former (if absolute or unconditional) is defined on a certain set E, the frequency is defined, for every sampling procedure s, on a finite subset E^* of E. (In short, $Pr: E \rightarrow [0,1]$, while $f: E^* \times S \rightarrow F$, where S is the set of sampling procedures and F the collection of fractions in the unit interval.) Hence it is not true that one gets a model or true interpretation of the probability calculus upon interpreting probability values as observed relative frequencies; at most we could say that we thus get a *quasi-model*. Fifth, if a stochastic theory (such as statistical mechanics, quantum mechanics, genetics, or some stochastic learning model) is construed as yielding frequencies, then there is no point in performing any measurements to check the theoretical formulas. (Likewise with all other physical concepts, for example the one of eigenvalue of an operator representative of a physical property: if eigenvalues were interpreted as measured values, as the orthodox school has it, then there would be no point in carrying out any actual measurements.) What makes both theory and experiment indispensable is that they are radically different: that a theory is not a summary of experiments, and that no run of experiments replaces a theory. It takes the two of them to engender a new item of knowledge.

In short, neither the subjectivist nor the dualist interpretation of probability have a place in theoretical physics: what do have a place in it are the following strict and objectivist interpretations: the *propensity* interpretation (Popper, 1959) and the *randomness* interpretation. On the former, a probability value is a measure of the strength of the tendency for something to happen: probability is just quantified potentiality, with reference to physical systems, whether simple or complex, free or under the action of other systems, and in particular whether under observation or not. (Actually this is my own version of the propensity interpretation, as found in Bunge (1967e). Popper's version (*ibid.*) concerns the object-experimental arrangement compound and could therefore be mistaken as supporting Bohr's thesis of the indextricable unity of the two – as has in fact been interpreted by Feyerabend (1968). In a personal communication Sir Karl has indicated agreement with my reinterpretation. See the illuminating discussion by Settle (1972).) On the second interpretation, probability is the odds or weight of an event belonging to a random collection (e.g., a Markov sequence) of events.

On either interpretation the probability of an event is an objective property of it: it is inherent in things; likewise a probability distribution is interpreted as an objective (but potential rather than actual) property of a physical system. The difference between the propensity and the randomness interpretations is that the former is wider, for it does not require the events to be random, while the randomness interpretation holds only for random events and therefore calls for criteria allowing one to find out whether the given set of events is a random one. In other words, the randomness interpretation of probability may be regarded as the restriction of the propensity interpretation to the subset of random events. On either interpretation the probability of, say, a transition from one state of a system to another state is just as objective as the speed of the transition: it is not in any way linked to ignorance, or to uncertainty, or conversely to the strength of our beliefs (which are usually too strong anyway). We shall call the two interpretations by the name *physical probability*.

Whether or not one is suspicious of the propensity concept, one must surely regard the probabilities occurring in physics as physical properties on a par with internal stress and electric field strength. The reason is this: all the independent variables of a probability function in a physical theory stand for physical systems or properties thereof. (Even time, the least tangible of all physical variables, can be elucidated in terms of events and frames: Bunge, 1968d.) There is no way of smuggling the observer and his mind into a theoretical probability statement by arguing, for example, that quantum mechanics does not concern autonomous systems but rather a complex constituted by a microsystem, an experimental arrangement (which one, pray?), and the operator of the latter. First, because this is simply wrong: most quantum-mechanical formulas are about microsystems embedded in a purely physical medium (which is very often absent).

This is not a matter for pronouncements *ex cathedra* but a matter of analysis of the formulas concerned, and this analysis will not be exhaustive unless the formulas are written out explicitly, i.e., in the axiomatic way that is so distasteful to the enemies of clarity. A second reason is that even those formulas which do concern an object-environment complex (e.g., a molecule immersed in an electric field), fail to be about an observer proper, i.e., a psycho-physical being. For, if they were, the quantum theory ought to enable us to predict not only the microsystem's behaviour but the observer's conduct as well, which unfortunately it does not. In conclusion, there is no ground for asserting that the cognitive subject enters theoretical physics, in particular the quantum theory, via probability and the state vector. And if he does not use these doors, it is difficult to see how else he could get in.

3. DISTINGUISHING APPARATUS FROM OBSERVER

3.1. *Approaches to Measurement Theory*

Many authors describe a measurement as an interaction between an object and an observer, or even as a synthesis of the two. But whereas some writers mean by "observer" a cognitive subject with his full psychical equipment, others mean a classically describable apparatus, and still others prefer to keep silent, hence ambiguous. If a difference between an observer and his equipment is not made, and if an observer is allotted a supra-physical mind (e.g., an immortal soul), then measurement becomes a gate through which soul and spirit flow not only into the making of physics, but also into the things themselves, which thereby cease to be things in themselves. Indeed, a standard argument against realism is from the nature of microphysical measurement. We must therefore take a look at the theory of the latter, or better at the various programmes for setting up a measurement theory, for there are several and none has been fulfilled. This we must do not only in the interest of epistemology but also in the interest of experimental physicists, for if they were indistinguishable from their equipments, then either they should be paid no salaries or they should be allotted no funds for the purchase and maintenance of experimental facilities.

Essentially the following approaches to measurement theory in relation to the quantum theory can be found in the literature.

(1) *Naive realism*: (a) basic measurements are direct, i.e., in no need of theories; (b) derivative or indirect measurements can be taken care of by the available physical theories supplemented with mathematical statistics; (c) upshot: no special measurement theories are required. *Criticism*: see the next point.

(2) *Critical realism*: (a) there are no direct precision measurements, particularly in microphysics; (b) any detailed theory of the measurement of a physical magnitude (e.g., time reckoning) or of the preparation of a physical system (e.g., a proton beam with a given velocity distribution) calls for a number of general theories as well as a definite model of the experimental equipment (e.g., a cyclotron theory is an application of classical electrodynamics or, if preferred, it is a piece of relativistic technology); (c) since measurements are specific and they involve macrophysical systems, genuine theories of measurement (unlike the phony ones found in some books on quantum mechanics) cannot help being specific and involving fragments of classical theories (e.g., classical mechanics and optics); (d) no adequate *general* theory of measurement is available, either in classical or in quantum physics, and moreover it is doubtful that any can be developed, precisely because there are no general measurements and every macrophysical event crosses several boundaries between the various chapters of physics. This is, of course, a thesis of the present book.

(3) *Naive operationism* (textbook philosophy): (a) every physical theory, in particular quantum mechanics, concerns actual or possible measurement operations and their outcomes; thus a hamiltonian operator represents an energy measurement and its eigenvalues are measurable energy values; (b) consequently there is no need for a special theory of measurement.

Criticism: (i) there is both a structural and a semantical difference between a theoretical magnitude and its experimental partner, if any (recall Section 2.2); (ii) if general theories did concern empirical observations, then one of the two – theories or observations – would be redundant and the choice of equipment should make no difference.

(4) *Radical operationism* (Ludwig, 1967): (a) basic measurements are direct; (b) a basic theory, such as quantum mechanics, should be concerned with basic measurements and be derived from an analysis of the physics of measurement.

Criticism: (i) there are no direct measurements, at least not on microsystems (see the above criticism of naive realism); (ii) scientific analyses, whether of concepts or of operations, far from being extrasystematic, are performed with the help of theories; (iii) in particular, an analysis of a measurement presupposes a number of theories, both substantive (e.g., electromagnetic theory) and methodological (particularly mathematical statistics).

(5) *Strict Copenhagen view* (Bohr, 1958a): (a) a measurement process is one in which object, apparatus, and observer become fused into a solid block, so that they lose their identities; (b) this unity is peculiar to the quantum phenomenon, which is thus unanalysable; (c) "the quantum-mechanical formalism permits well-defined applications referring only to such closed phenomena" (Bohr, 1958a, p. 73); (d) a theory of measurement would attempt to analyse such a unity, distinguishing between subject and object and finding out the precise form of their interaction, thus destroying the irreducibility and irrationality that characterises quantum phenomena; (e) consequently no attempt to build a quantum theory of measurement should be made (Rosenfeld 1964).

Criticism: (i) although a measurement act does involve an observer (and a number of other things as well), physics is not about sentient beings but about physical systems, sometimes under control but most often free and in any case devoid of mental components; (ii) it would be desirable to have a number of genuine detailed quantum theories of real (hence specific) measurement processes, theories capable of explaining and predicting the whole chain starting from an elementary event (e.g., a photochemical reaction) and ending up in an observable macroevent (e.g., the blackening of a photographic plate): to wish otherwise is sheer obscurantism.

(6) *Von Neumann's view* (von Neumann, 1932, 1955): (a) a measurement process is an object-subject interaction characterised by the arbitrariness of the frontier between the two (i.e., a cut can be made for the purpose of analysis but its position is conventional); (b) rather than being an application of quantum mechanics and other physical theories, a quantum theory of measurement requires suspending the main postulate of the latter (Schrödinger's equation or its equivalent), adopting in its stead the projection postulate, according to which the measurement of an observable throws the state vector onto any of the eigenvectors of the observable concerned; (c) the ensuing theory of measurement is quite general and

moreover it gives quantum mechanics its operational meaning. Since this view is supposed to be the standard one, we shall concentrate our attention on it.

3.2. *The Standard Account of Measurement*

The usually accepted account of the measurement process is the one given by von Neumann in a book (von Neumann, 1932, 1955) that passes almost universally, though wrongly, for offering an axiomatic and consistent formulation of quantum mechanics. This seems to have been the first time the observer was systematically allotted a prominent role in the account of experimental arrangements. Von Neumann made it clear that by an observer he meant not just a measuring apparatus but a human subject capable of "subjective apperception" (von Neumann, 1932, p. 223). He even thought it necessary to rope in the doctrine of psychophysical parallelism. Von Neumann also insisted (von Neumann, 1932, p. 224) that the frontier or cut between observer and observed system can be displaced at will. More precisely, he proposed dividing the world into three parts: the observed thing I, the measuring apparatus II, and the observer III. The frontier, he claimed, may be traced either between I and the compound system II + III, or between the physical complex I + II and the psychophysical entity III. In either case (a) a measurement is regarded as being something very different from, say, the action of an external magnetic field on a spinning microsystem – precisely because of the unpredictable, nay capricious, intervention of the conscious mind, and (b) the measurement process is neither controllable nor fully reducible to physics, for it involves subjective apperception and arbitrary choice (von Neumann, 1932, pp. 223ff.).

The active role this account assigns to the conscious observer in determining the outcome of a measurement is best brought home by the following imaginary procedure, which may be called the *mensura interrupta* technique. You set up an experimental arrangement to measure a given magnitude on an object of a certain kind and operate the device all the way but abstain from taking the final reading. After a while you flip a coin: if heads you look at the pointer and register its position; if tails you walk out of the laboratory. Being a subjectivist, you are unwilling to distinguish the *physical* fact that the pointer came to rest at a given position, from the *psychical* fact that you take or do not take cognizance of such a physical fact: what is more, you refuse to believe that there is such a thing as an autonomous physical fact. Then you are bound to conclude that the outcome of a measurement, i.e., the value of the magnitude concerned, depends on the observer's consciousness. Assume further that you abide by the operationist tenet that calculated values are possible measurement values: then you will conclude that the conscious observer is an essential member of the quantum theory and, in general, that Man can no longer be ignored by physics (Heitler, 1963, pp. 34–35).

To return to von Neumann's tripartition of the world: inconsistently enough this division, claimed to be essential, is not embodied into a theory: it is vacuous. In fact, *nowhere* in von Neumann's book are the properties of the observer (system III) specified, even sketchily: (a) his discussion of compound systems (von Neumann, 1932, Chapter VI, Section 2) which sets the stage for his treatment of the measurement process (von Neumann, 1932, Chapter VI, Section 3) concerns the "observed" object coupled to the measuring apparatus, i.e., I+II, a compoud of physical systems with no admixture of mental components; (b) von Neumann says explicitly that the subject "remains outside the calculation" (von Neumann, 1932, pp. 224, 234). Now, something that does not occur in the theory, yet is supposed to be its distinctive mark (as opposed to a classical measurement theory), is a phony item, a ghost, a hidden variable in the bad sense of this expression.

But the cognitive subject is not the only ghost in von Neumann's theory, or rather pseudotheory, of measurement. An actual ingredient of it has a ghost-like quality as well: this is the state of the observed system before a measurement is actually performed. For, if this state is empirically unknown and moreover unknowable, then it should not occur in a theory that is supposed to abide by an empiricist philosophy. (On the other hand it can occur on any alternative philosophy, for it may be regarded as a hypothesis to be tested by observation.) Furthermore to maintain, as von Neumann did, that a measurement brings about a transition from that unknown state into an unpredictable eigenvector of the measured "observable", is to explain the obscure by the more obscure.

In any case, a sketch of a theory of highly idealised measurements of arbitrary magnitudes, surrounded by empty talk about idle observers, cannot pass for a theory of actual measurement even though it be approved of (but never used by) the bulk of the physical profession. A reason for the failure of von Neumann to provide a genuine theory of measurement is that there is no such thing as an arbitrary measurement. A second reason is that he accepted uncritically the orthodox interpretation of quantum mechanics he learned from physicists, without realising that this interpretation renders measurement redundant. (Recall the strict Copenhagen view discussed in Section 3.1.) Indeed, according to that interpretation, an eigenvalue is not a value actually possessed by a system but rather a measured value. (We have argued in Section 2.2 that this interpretation is adventitious and invalid.) Hence no separate theory of measurement should be necessary if the orthodox interpretation is adopted. Now, if eigenvalues are measured values, then eigenfunctions must represent states of systems under observation. On the other hand a general state vector (a linear combination of eigenfunctions or eigenvectors) must represent a state of a system before or after it is observed, particularly if the subjective interpretation of probability is embraced, as von Neumann did half of the time. He did not see that there is no point in building a whole theory (quantum mechanics minus measurement theory) centered around the equation of evolution of such unobservable states. Nor did he realise that the duality of his two kinds of processes, the one of collapse of the state vector upon measurement (process 1) and the other of smooth ("causal" in the incorrect standard terminology) evolution, in accordance with the Schrödinger equation (process 2), contradicts the very philosophy he espoused, for one does not write out a whole theory concerning a process that is in principle unobservable. Finally, von Neumann did not see either that – as Margenau pointed out long ago (Margenau, 1936) – all the actual calculations in quantum mechanics, in particular those which have been checked by experiment, concern processes of the second kind, namely those satisfying the Schrödinger equation, rather than processes of the first kind. Therefore, if a general quantum theory of measurement were possible, which is doubtful, the natural thing to do would be to drop von Neumann's projection postulate and apply Schrödinger's equation (or an equivalent) to the object-apparatus complex regarded as a purely physical two-system entity (Everett, 1957; Wheeler, 1957; Daneri et al., 1962; Bohm and Bub, 1966; Bunge, 1967a) – or, even better, to treat it as a many-body problem. In any case a measurement theory would be an application of a basic theory rather than a chapter of it. However, the very possibility of a general theory of measurement, whether classical or quantal, is problematic, because a universal meter would measure nothing in particular.

So we have this anomalous situation. First, it is claimed that only a discussion of empirical operations, such as measurements, can supply the content or physical meaning of the mathematical formalism of the quantum theory. This tallies with the obsolete verification doctrine of meaning but is inconsistent with the practice of designing, analysing, and evaluating empirical operations in the light of theories. Second, the standard quantum-mechanical theory of measurement (von Neumann's) does not have the blessing of the proponents of the equally standard interpretation of quantum mechanics. Third, von Neumann's theory of measurement is practically nonexistent and it is supposed to contain a concept, the one of observer, that is extraphysical and moreover has not been incorporated into the (pseudo) theory: it remains outside the formulas of the latter, it hovers above them without getting actually mixed with the actual components of the theory. Fourth, no realistic cases have been handled with the help of von Neumann's theory of measurement. He himself gave a single example which, for being concerned with two mass-points, is not an example of a real measurement; he left the discussion of realistic, hence enormously complicated, examples to the reader (von Neumann, 1932, p. 237). As a consequence this theory remains *untested*: indeed, it has failed to yield a single verified prediction. Even a distinguished defender of orthodoxy grants that "there is no realistic theory of actual measuring devices" (Stapp, 1971).

In short, the standard quantum theory of measurement, which is alleged to enthrone the observer in theoretical physics, is a ghostly one altogether. Consequently the usual attempts to discuss the foundations of quantum mechanics, and in particular its meaning, in terms of the theory of measurement, are as ill-advised as the attempts to disclose the nature of man via theology. Worse, the point of measuring is to get down to particulars, which can only be done with the help of specific pieces of equipment. And any such particular measuring set-up calls for a specific theory. And any such specific theory is an application of a number of general theories: actually it is a set of general theories together with a definite model of the experimental situation. Hence no single theory can be expected to account for every possible measurement device, except in such a superficial way that it will be helpless to explain and predict the behaviour of a single particular experimental arrangement. Therefore the strict Copenhagen view, according to which no time should be wasted in trying to build a quantum theory of measurement, is right though for the wrong reason. But no matter what stand one may take on this controversial issue, the philosophically important point is that no existing quantum theory of measurement is concerned with The Observer, *pace* the repeated verbal attempts to smuggle Him into the picture.

3.3. *Experiment Presupposes Realism and Confirms It*

Strange as it may seem, the opponents of realism try to argue from the most tangible aspects of physics, namely laboratory physics. The favorite arguments are these. "A physical quantity has no value unless it is measured; now, measurement is a human action; hence physical quantities acquire a precise value only as a result of certain human actions. Likewise, a thing is in no definite state unless it is prepared to be in a given state; now, a state preparation is a human action; hence physical systems adopt definite states only as a result of certain human actions."

These arguments, though popular, are circular, for their conclusions assert the same thing as their major premises. In fact, 'to measure' and 'to prepare' are pragmatic terms which the minor premises spell out. The major premises say all the nonrealist wishes to assert, namely that whatever is, is so because somebody has decided to make it that way or, equivalently, that properties and states have no autonomous existence but are observer-dependent. Moreover these premises are false, for they rest on a confusion between being and knowing. Surely a magnitude has no *known* value unless it is measured. But this does not entail that it *has* no definite value as long as it is not being measured. The contrary thesis amounts to the claim that the researcher does not investigate the world but creates it as he proceeds, which is philosophically ludicrous, as it leads to subjective idealism and ultimately to solipsism.

The nonrealist thesis is also mathematically untenable. In fact, when formulating a physical theory one will state, for example, that a certain property is represented by a real-valued function, and one will assume or hope that measurements will be able to sample such values at least within an interval of the whole range of the function. One assumes, in other words, that the function *has* certain values all the time for, if it did not have them, it would not be a function – by definition of "function." Similarly with the operators assumed to represent dynamical variables: they are supposed to have definite eigenvalues even while no measurement of such properties is being performed, for otherwise they would not be well-defined mathematical objects. This does not entail that a physical system has at all times a sharp position and a sharp velocity (or, in general, that it is at every instant in a simultaneous eigenstate of all its "observables"), only we do not happen to know such precise values. Since in quantum mechanics the dynamical variables are random variables, they have definite distributions (even for a single physical system) rather than definite numerical values. But these distributions and, in general, the bilinear forms built with the operators and the state vectors, do have definite values at each point in space and time, for they are ordinary point functions.

In short, the thesis that the values of functions and the eigenvalues of operators are measured values is mathematically untenable. Surely the decision to measure or to prepare a system, as well as the ensuing laboratory operations, are the doings of humans, and the outcomes of these actions will depend on them as much as the outcome of any other human action. But humans are part of nature, their action on their environment is efficient only insofar as it is based on some knowledge of nature, and only the physical aspect of such actions is relevant to physics: minds have no direct action on things and, even if they had, physics would not be competent to account for them. Surely the act of preparation modifies the initial state of the thing, whether or not it is a microsystem; but for such a change to occur the thing must be available or it must be produced out of other things that were there to begin with; also, the change must be a thoroughly real one even when steered by a subject.

With the exception of extreme subjectivists, who hope to get away without any empirical operations, everyone agrees that measurement and experiment are essential to physical research. Now, in order for any such operation to furnish genuine empirical evidence, it must be real: dreams and gedanken-experiments can be heuristically valuable but they prove and disprove nothing. In other words, the least one will do when assessing an experiment is to ascertain whether the experimental set-up is actual, otherwise one will speak of a plan for an experiment or even of a fraud. Of course, any experimental arrangement is artificial in the sense that it is planned, made, and controlled by humans, either directly or indirectly. But so is a car and so is an artificial satellite, and yet nobody would

mistake such artifacts for observers. Now, we cannot satisfy ourselves
that a certain experimental set-up is real unless its immediate environment
is actual as well, for otherwise there would be no point in constructing
insulators and in making temperature and pressure corrections, in inspect-
ing the system for external disturbances and leakages, etc. Furthermore,
every component of the system must be real for the whole to be real.
If the components of a complex system were mental rather than physical,
they would give rise to a psychical whole. This contradicts the claim of the
Copenhagen philosophers that, while macrosystems (e.g., apparatus)
may be real, their atomic constituents lack autonomous existence. Of
course one often makes the mistake of believing that something is out
there while it is actually missing. But errors of this kind can eventually be
recognised as such, and such corrections show how much store we do set
by the assumption that in the laboratory one handles real things.

   In short, experimental physics assumes the reality of the objects it
manipulates, and it tests some of the theoretical hypothesis made about
the existence of physical systems. Experimental physics has no use for a
physical theory that makes no existence assumptions, and theoretical
physics can expect no help from experimentalists who are not willing to
soil their hands with real things.

4. FOUR POSSIBLE STYLES OF THEORISING

4.1. *The Realist and the Subjectivist Versions*

In order to better assess the merits and demerits of the various philosophies
discussed so far, we shall try to formulate in a cogent way (i.e., axiomatically) a very simple theory in four different guises, each corresponding
to one of those philosophies. (This will have the side effect of underpinning
the thesis that scientific research is far from being philosophically neutral.)
We shall start with the realist and the subjectivist theories, which can be
dealt with jointly because of their unambiguous monistic character.

   Let the theory concern a physical system (alternatively, a subject)
which either is in one of two states named A and B, or jumps from one
of them to the other in such a way that each of the four possible events,
$\langle A, A \rangle$, $\langle A, B \rangle$, $\langle B, A \rangle$, and $\langle B, B \rangle$, has a definite probability. (The first
and the fourth are, of course, null events.) Five specific primitive (unde-
fined) concepts will do the job: the set Σ of systems (alternatively, of

subjects), a state function S, two constants A and B, and the probability function Pr. The difference between the two theories, the realist and the subjectivist one, lies in the referent: in the former case the reference class Σ is interpreted as a set of physical systems, while in the subjectivist case it is interpreted as a set of subjects. Accordingly the functions S and Pr become either properties of a physical system or properties of a subject. To save space the subjectivist interpretation will be indicated in parentheses and in *italics*. Only the axiomatic foundations will be laid down.

Axiom 1. There are physical systems (*subjects*) of the kind Σ. [In a slightly more detailed way: (a) $\Sigma \neq \emptyset$. (b) Every $\sigma \in \Sigma$ is a physical system (*subject*).]

Axiom 2. Any physical system (*subject*) of the kind Σ is in either of two states (*states of mind*): A and B. [More explicitly: (a) S is a many-to-one function from Σ into $\{A, B\}$. (b) A and B represent states (*states of mind*) of a physical system (*subject*) of the kind Σ.]

Axiom 3. (a) Pr is a probability measure on $\{A, B\}^2$. (b) The probability of any pair in $\{A, B\}^2$ is nonvanishing [all transitions are possible]. (c) $Pr(\langle A, A \rangle) + Pr(\langle B, B \rangle) = 1$. (d) $Pr(\langle A, B \rangle)$ represents the strength of the tendency of propensity (*rational belief* or *certainty*) with which a physical system (*subject*) in state (*state of mind*) A jumps into state (*state of mind*) B, and similarly for the other probability values.

The ostensive differences between the two theories are these. (a) While the realist theory concerns an idealised physical system (a model of plenty of real situations), the subjectivist theory concerns an idealised subject (hardly a suitable model of anyone save an extreme moron). (b) Whereas the realist theory informs about physical events, the subjectivist one informs about psychical events. (c) While the realist theory involves transition probabilities that can be checked by observing frequencies of external events, the subjectivist theory involves introspectively observable transition frequencies. (d) While the realist theory is testable in a physical laboratory, the subjectivist one is not testable in this way.

Both theories are phenomenological or black box theories in the sense that neither accounts for the transition mechanism. But they can be deepened so as to explain the transitions. In either case such a deepening calls for the introduction of new basic concepts and correspondingly of new postulates. (Remember the unspoken rule: For every new primitive, at least one new formal and one new semantical postulate.) Thus the realist theory can be expanded into a stronger theory explaining the transition probabilities in terms of, say, the state occupation number. For example, the probability of the event $\langle A, B \rangle$ could be set proportional to the occupation number of the state A and inversely proportional to the occupation number of B. Or else a hidden variables theory might be set up: a theory containing further variables and equations of evolution for them that would explain both the existence of the states and the transitions between them. Any such stronger theory would still be a physical theory. On the other hand the subjectivist theory might be expanded in either of the following opposite directions: the new variables could be further psychological concepts, or some of them could be neurological (physiological) concepts. In the first case a homogeneous extension would be obtained: the new theory would remain within psychology. But in the second case the stronger and deeper theory would have a mixed character: it would contain both psychological and physical (or rather neurophysiological) variables, so that it would describe a two-level system. A still further extension might be able to analyse every psychological variable remaining in the former extension, in neurophysiological terms. Let us risk the following conclusions: any deepening of a realist theory retains its physical character, while some attempts to deepen a subjectivist theory change its character, thus defeating the philosophy of subjectivism. In other words, it would seem that subjectivism can be kept at the price of avoiding further deepening, which is not the case of realism. But we are not now concerned with depth. Our aim was just to show that a theory can be cast either in realist or in subjectivist terms. We shall presently see that none of the other two philosophies we have been discussing allows this.

4.2. *The Copenhagen Predicament*

In a theory built in the pure Copenhagen style there should be a single reference class: the set of sealed units constituted by the object, the observation set-up, and the observer. At first sight there should be no difficulty in obtaining the Copenhagen version of any given physical theory, such as the one expounded in the last section: it would seem that a reinterpretation of Σ as the set of trinities should suffice. As a matter of fact there are two technical obstacles in the way, one of a formal nature, the other semantical.

The mathematical obstacle to the Copenhagenisation of theories is this. The claim that the referent of a theory is single (equivalently, that its reference class is homogeneous in the sense of Section 1.1), and furthermore unanalysable, contradicts the claim that every"quantity"(magnitude) is relational in the sense that it concerns not just the physical system of interest (e.g., an atom) but also its (artificial) environment and the observer in charge of the latter. These two claims of the Copenhagen school are obviously mutually contradictory, for the first boils down to the assertion that the domain of the functions (e.g., probability distributions) concerned involves a homogeneous set of indivisible blocks, while the second claim boils down to the assertion that that domain involves the cartesian product of the set of physical systems by the set of apparatus and the set of observers. So much for the mathematical difficulty.

The refusal to analyse the referent *unum et trinum* renders the interpretation task hopeless, for the properties to be assigned to that referent are neither here nor there: they are neither strictly physical nor strictly psychological. This is why the Copenhagen doctrine is as obscure as the doctrine of the trinity, according to which the Father (Apparatus), the Son (Microsystem), and the Holy Ghost (Observer) are united in one Godhead (Quantum phenomenon). Take, for instance, the notion of state occurring in the microtheory expounded in the previous subsection. While in the realist (alternatively the subjectivist) interpretation A and B stand for physical states (alternatively mental states) of a system of a definite kind (physical or psychical), in the Copenhagen interpretation they should represent total or psychophysical states of the block: system-apparatus-observer. But no existing science accounts for such complex (yet unitary) entities.

In conclusion, it is impossible to build a *consistent* theory in the Copenhagen style. In other words, the Copenhagen interpretation of the quantum theory is inconsistent, and moreover incurably so (see also Chapters 5 and 6). Fortunately the baby – quantum mechanics – need not be thrown away together with the bath water: there are consistent alternative formulations of the theory (e.g., Bunge, 1967a).

4.3. *The Dualist Version*

Let us return to the imaginary microtheory discussed in Section 4.1. Its axiomatic reformulation in a dualist (e.g., operationist) spirit would require two further distinct sets: the set I of instruments and the set O of observers or operators. These various items should be regarded as interacting but also as distinct. (If they were indistinguishable, if they constituted a solid block, they could hardly interact.) Hence the corresponding concepts must be taken as mutually independent primitives. The dualist version of our microtheory would then be based on seven rather than five undefined concepts.

Now for an axiom system to be satisfactory, it must contain axioms specifying both the mathematical structure and the factual meaning of every one of its basic technical terms. (This may be called the condition of primitive completeness: see Chapter 6, Section 4.) This is well-nigh impracticable in the case of the additional primitives I and O, and even if it were feasible it would be hardly desirable. It is impracticable because, while Σ is handled by a strictly physical theory and moreover a well-defined one, I and O require going far beyond that theory. In fact, the characterisation of any apparatus in theoretical terms calls for a whole assembly of fragments of different theories. Likewise, the specification of any observer would take all the sciences of man: anthropology, psychology, sociology, etc. The theory would then acquire a gigantic size in case it could be developed at all. The dualist programme is therefore unfeasible. It is not desirable either, for the following reasons. First, it would render general theories impossible, for a general theory is one that is not tied to any special kind of experimental set-ups. Second, the dualist program would render the progress of physics dependent on the state of the sciences of man – whence, if adhered to at the end of the Renaissance, physics would not have taken off. After all, modern physical science was born in opposition to anthropocentrism.

In conclusion, of the four conceivable types of theorising two are impracticable: the Copenhagen and the dualist ones. The realist and the subjectivist approaches are feasible but only the former yields objective, testable, and in principle improvable theories.

5. CONCLUSION: REALISM UPHELD

We started by distinguishing two kinds of interpretation of physical symbols: strict interpretation, which matches the mathematical structure of the corresponding idea, and adventitious interpretation, which overflows it. We have shown that, while in theoretical physics only strict interpretations are warranted, adventitious interpretations (e.g., in terms of operations) are called for in experimental physics, but they are valid only insofar as they are backed by theories (e.g., theories accounting for the operations).

We then applied the previous distinction to some basic physical concepts. The upshot was that the only strict interpretations in theoretical physics are either realist or subjectivist, all others being adventitious. But we showed that there is no ground for the subjectivist interpretation of two functions that pass for being the gates through which the mind enters the physical picture, namely the state vector and probability. This we did by examining the independent variables, i.e., the domains of those functions, as well as by recalling some of the presuppositions and goals of scientific research. The discarding of subjectivism left us with realism as the sole viable philosophy of physics.

Next we explored the possibility of casting one and the same theory in each of the four competing philosophical molds: realism, subjectivism, the Copenhagen view, and dualism (in particular operationism). It turned out that, while the first two projects are viable, the subjectivist one does not lend itself as easily to generalisation and deepening, and in any case is incurably untestable, hence nonscientific. As to the Copenhagen version, it proved impossible without contradiction, and the dualist (in particular operationist) formulation proved impracticable. Once again realism was vindicated as the sole realistic philosophy of physics.

Finally we turned our attention to measurement theory, often said to be another door through which the spirit enters our new picture of the world. We found that the standard theory (von Neumann's) is ghostly on more than one count: it hardly exists as a realistic theory of actual measurements and it talks about an observer that is supernumerary, as it occurs nowhere in the formulas. Here again, our analysis has upheld realism and, in particular, the trite yet important thesis that physics is about physical systems – notwithstanding the nonrealist phraseology that so often surrounds physical formulas and physical operations.

Now, there are a number of realist views (hardly theories) of knowledge. Which one does our semantical and methodological analysis support? The answer is, of course, *critical realism*. This view is characterized by the following theses:

(1) There are things in themselves, i.e., objects the existence of which does not depend on our mind. (Note that the quantifier is existential, not universal: artifacts obviously depend on minds.)

(2) Things in themselves are knowable, though partially and by successive approximations rather than exhaustively and at one stroke.

(3) Knowledge of a thing in itself is attained jointly by theory and experiment, none of which can pronounce final verdicts on anything.

(4) This knowledge (factual knowledge) is hypothetical rather than apodictic, hence it is corrigible and not final: while the philosophical hypotheses that there are things out there, and that they can be known, constitute presuppositions of scientific research, any scientific hypothesis about the existence of a special kind of object, its properties, or laws, is corrigible.

(5) Knowledge of a thing in itself, far from being direct and pictorial, is roundabout and symbolic.

Critical realism is the most fertile of all epistemologies because it encourages us to look beyond every theory, however successful and therefore perfect it may look at any given time. In particular, it encourages the exploration of new pathways in fundamental physics – which, every one seems to agree, can use some radically new ideas. Critical realism also encourages the reconstruction of the extant theories in a clearer and more cogent way – as will be seen in the next chapter.

NOTE

* Some paragraphs are reproduced from Bunge (1969b) with the permission of the editor and publisher.

CHAPTER 5

QUANTUM MECHANICS
IN SEARCH OF ITS REFERENT

Quantum mechanics (henceforth QM), probably the most powerful of all scientific theories, is also the one with the weakest philosophy. This weakness resides mainly in the inability to state unambiguously and persuasively what the genuine referents of the theory are. And this inability derives from a desire to comply with a philosophy that wavers between undiluted subjectivism and straight realism. Indeed, the usual interpretation of QM, as found e.g. in the classical treatises of von Neumann (1932) and Dirac (1958), as well as in standard textbooks such as those of Bohm (1951) and Landau and Lifshitz (1958), has been cast in the spirit and letter of the early logical positivism fashionable among scientists between the two wars (see Frank, 1938, 1946, von Mises, 1951; and Reichenbach, 1951).

The commitment of the usual formulation of QM to a moth-eaten philosophy, that hardly any living philosopher holds any longer, is largely responsible for the interpretational inconsistencies and obscurities of the theory. (We are not concerned with the mathematical inconsistencies such as the infinities.) Much of this confusion is acutely felt by the beginner, but the practitioner learns to live with it. The latter gets used, in fact, to handling a conceptual instrument he does not profess to understand: occasionally he goes as far as claiming that the lust for understanding is a sinful remainder from classical physics. He may admit that QM is befogged and sometimes he makes a virtue out of this, arguing that quantum events are ultimately opaque to reason (cf. Bohr, 1934) and that we must consider ourselves lucky if, without understanding in the classical sense of the word, we succeed in computing predictions borne out by observation and experiment.

This situation is intolerable to the philosopher and the historian of science who realise that QM is a triumph of reason, that nothing is crystal clear at the start, and that the barriers to reason are in the habit of crumbling down one after the other. The philosopher may suspect that the fog enveloping QM is of a philosophical nature and can therefore be pierced through with tools that are not to be found in the standard tool kit of the physicist – namely, logic, semantics, epistemology, and methodology, Furthermore, the philosopher may suspect that the fog surrounding QM has been delaying progress in basic physical theory during the last thirty years, i.e., after the main edifice of QM was erected. Indeed, the successful applications of basic QM have been so numerous that only a minority of physicists are exploring radically new paths. Theoretical physicists have become, in this respect, even more conservative than theologians. As a result no breakthrough has been achieved in recent times in basic microphysical theory nor will any be attempted as long as the present theory continues to be regarded as perfect or nearly so: complacency leads to stagnation and decadence. Only a Quantum Concile can help us out of it.

It is therefore advisable to sketch the philosophical fog that prevents us from seeing ahead, and to free QM from it. The performance of these two tasks, the critical and the one of reconstruction, should be not only philosophically interesting but also useful to the advancement of knowledge.

1. FACING THE FOG

Like every other physical theory, QM consists of a mathematical formalism endowed with a certain interpretation. The usual interpretation of QM is known as the *Copenhagen doctrine* and was worked out by some of the giants who built the theory: Bohr, Heisenberg, Born, Dirac, Pauli, and von Neumann. This doctrine, or rather set of doctrines, is well known to physicists. What most of them do not seem to realise is that the Copenhagen doctrine is scientifically and philosophically untenable because it is inconsistent and not thoroughly physical. Let us take a quick glance at these two fatal traits of the orthodox doctrine.

The inconsistencies of orthodox QM are both formal and semantical, and they can be found both in the body of the standard theory and in the metatheory. A typical inconsistency of the formal type is this. On the one hand it holds, and rightly so, that most microproperties are peculiarly quantum-mechanical, *i.e.* nonclassical – which accounts for the novel character of QM *vis à vis* classical physics. But on the other hand these properties are said to characterise human manipulations (laboratory operations) rather than bits of matter. Now such operations take place on the macrophysical level and are therefore classically describable. In short, the doctrine contains the contradictory metastatement: "The quantum-mechanical symbols concern nonquantal (classical) facts".

The source of this contradiction is a philosophical one: it originates in the tenet that physical theory is not about reality (a supposedly metaphysical misfit) but about human experience (a supposedly crystal clear thing). What is true is, of course, that physical theory is *about* reality and is *tested* through human experience, by contrasting some of the logical consequences of the theory with facts out there and under experimental control. The Copenhagen doctrine specifies that bit of empiricist philosophy in the following way: "There are no autonomous quantum events but only observer-dependent quantum items: the observation or measurement operations generate the entities in given states." But this statement is inconsistent with the very practice of the quantum physicist: indeed, most of the problems we handle within the context of QM concern physical or chemical systems that, by hypothesis, do not interact with pieces of apparatus. Moreover, the quantum theory of measurement is virtually nonexistant, as it is incapable of accounting for the specific traits of measuring instruments that make measurements possible. (Recall Chapter 4, Section 3.) Furthermore the tenet in question is inconsistent with the assumption that at least the observer himself is real and is composed of microsystems. Indeed, if every one of the atoms in my body exists only insofar as I can observe it, then I – a system of atoms – do not exist, if only because I have other things to do than to unceasingly observe my microphysical constituents. In short, the Copenhagen doctrine is logically inconsistent, and this blemish derives from its adopting a subjectivist philosophy. (For further inconsistencies see Bunge, 1959b, 1967a; Landé, 1965, and Popper, 1967.)

The doctrine is inconsistent in a further sense as well, namely semantically. Let us call a theory *semantically inconsistent* if, at some point, it lets in predicates that are not kindred to the basic predicates (primitive concepts) of the theory (Bunge, 1966 and 1967c). This can always be done, in a nonaxiomatic context, thanks to the logical law "If p, then p or q". In fact, if the formula p is asserted then "p or q" can be concluded even if q contains predicates that are totally alien to those involved in the premise p. Thus upon asserting Schrödinger's equation we could deduce "Either Schrödinger's equation holds or it is not the case that the observer creates the world," which is true and moreover equivalent to "If the observer creates the world then Schrödinger's equation holds." Of course this is cheating: the predicates "observer", "creates" and "world" were not included in the predicate base of the original dicourse: they come out of the blue.

Precisely this manoeuvre is constantly performed in the context of the Copenhagen doctrine. Example: State the problem of computing the possible energy levels of an *isolated* atom of a given species, and end up by interpreting the results of your calculation as the possible values obtained by an experimenter actively disturbing the atom – even though the atom is *ex hypothesi* so far away that no experimenter could possibly establish an interaction with it. This is going to fetch atoms and return with observers. Semantical inconsistencies like this one are indulged in all the time and they too originate in the adherence to a subjectivist epistemology, in particular operationism.

There is only one way to avoid such semantical inconsistencies, namely by fixing the primitive base (set of undefined concepts) and sticking to it – in other words, by axiomatising the theory. If the concept of observer is included among the basic predicates, then no semantical inconsistencies of that type need arise. But then syntactical inconsistencies are bound to arise as soon as one wishes to construct the observer out of microsystems that are observer-dependent. The way to avoid both kinds of inconsistencies is clearly to axiomatise the theory and to do it without using nonphysical predicates: i.e., to reformulate QM in an *orderly and strictly physical* fashion.

The formal and semantical inconsistencies that plague the standard formulations of QM could not be avoided by taking mild measures, for they originate in a dogmatic attachment to a philosophy that is inconsistent with the very goal of physical science: you cannot have a fully physical theory if it is to satisfy nonphysical requirements such as the postulate that there are no autonomous (observer-free) physical entities and properties thereof. The *semiphysical* character of the standard formulations of QM is obvious from its terminology: a symbol representing a physical property is called an *observable*, and a macrophysical system such as a reference system or a measurement apparatus is called an *observer*. Instead of speaking simply about a property of a physical system, the adherents to the Copenhagen doctrine will try to speak of an observable *tout court*, or of an observable whose numerical values are determined (or even defined!) by a certain sequence of laboratory operations. Sheer anthropocentrism is thereby indulged in.

Yet an analysis of the symbols occurring in QM belies this interpretation. Thus the position operator of the i th microsystem in a given aggregate will be written x_i: the subscript i names a concrete (but arbitrary) physical individual which, for all we know, may dwell by itself in a forlorn corner of the universe. And the quantum-mechanical average of x_i for a fixed individual i is a function of time, not a function of the observer and the parameters describing the circumstances of his observation acts. Something similar happens with every other "observable" in basic QM. In short, the observer is supernumerary in QM: it is introduced only to comply with the adopted philosophy but it is never taken seriously in the calculations.

Worse: the philosophy inherent in orthodox QM renders physics proper impossible by subordinating it to the psychophysiology of the human observer. Mind, it is not just that the statements of every physical theory ought to be empirically testable – which is all right. What the Copenhagen doctrine claims is that all those statements should *refer* to test situations, for otherwise they would be meaningless. (Some go as far as claiming that the observer's mind should be included as well: see Wigner, 1962 and Heitler, 1963.) What happens is that the Copenhagen school mistakes the *referent* of a theory for its *test*: it identifies a methodological issue with a semantical one. It therefore drags one of the muddles responsible for operationism. (Recall Chapter 1.)

The Copenhagen school, then, by smuggling the observer into QM renders the latter psychophysical rather than purely physical. This would have satisfied Mach and his offspring, the Vienna Circle, who had proposed unifying science on the basis of human psychology. But it fails to satisfy us, if only for the following reasons. Firstly because, as long as classical physics coexists with quantum physics, two mutually incompatible epistemologies would be used: a realist one concerning the macrolevel and a subjectivist one associated with the microlevel. Secondly because if the observer, with his full psychophysical equipment, were to enter physics as a referent, then physical theories could not possibly be subjected to tests without the assistance of a highly developed science of psychophysiology. But QM contains no single assumption concerning the constitution and behaviour of The Observer: not even the orthodox formulation specifies them. Yet, since the usual formulation does involve observers as referents, not just as builders and testers of the theory, then in order for the word 'observer' to make sense, a substantial body of psychophysiology ought to be adjoined to physics. As a matter of fact the converse process is taking place: namely, psychophysiology is using more and more physics and chemistry, whereas theoretical physicists who pay only lip service to the Copenhagen interpretation are succeeding in explaining and predicting physical facts without using psychophysiology. This shows that the observer concept is not only foreign to physical theory, but that it should be possible to reformulate QM without the help of this psychophysical concept. Let us perform a preliminary exploration of this possibility.

That a translation of the semiphysical statements characterising the Copenhagen doctrine into physical statements can be done in every case, is suggested by the following examples. The phrase

> *Event x appears to observer y*

when purged of its pragmatic ingredients is reduced to:

> *Event x occurs in the reference frame y* [which may or may not be inhabited by an observer].

And the expression

> *The uncertainty concerning the statement that event x will occur is y*

reduces to

> *The probability of event x is* $1-y$.

Notice that a translatability of this kind does not amount to a logical equivalence: in most cases the events handled by the theory are supposed to occur without the help of the cognitive subject, and a statement concerning the objective probability of an event differs, conceptually as well as numerically, from a metastatement regarding the probability assigned by someone to an object statement. The point is that such a translation is possible and that it must be performed if we wish to retain the distinction between the external world and the internal world. Let us then turn to a purely physical interpretation of the mathematical formalism of QM, even risking the rebuke of those (e.g. Rosenfeld, 1953, 1961) who believe that the mere raising of any doubts concerning the correctness of the basic principles of the orthodox version of QM is futile. To a philosopher futility – and also vanity – inhere in dogma not in doubt.

2. Lifting the fog

The standard formalism of QM can be interpreted in a strictly physical, in particular nonpsychological, way. In other words, it is possible to re-interpret QM just in the same way as classical physics is interpreted, i.e. assuming that the entities referred to by the theory – electrons, atoms, molecules, etc. – exist by themselves. This does not exclude, of course, the possibility for an experimenter to modify these things, e.g., by filtering out certain states, or even to show us that certain microsystems were but imaginary. Only, the experimenter will have to use physical means to this end: it won't do just to sit down, compute and invoke the Copenhagen spirit. In other words, if the experimenter is involved at all, it is *qua* entity capable of influencing physical events by physical means, either directly with his bodily motions or indirectly through automatic devices. The physicist's mind figures out the formulas involved in the theoretical predictions and eventually also in the design and interpretation of the experiment, but it does not act directly on the physical events under study and is therefore not the concern of the theory itself.

The recipe for building strictly physical versions of QM is this: "Take the standard formulation, purge it from its subjectivistic elements, and finally reorganise logically what is left." The subjectivist elements are, of course, the observer concept and all the notions related to it, particularly those of observable and subjective probability. In the usual formulations of QM, the concept of observer occurs, e.g., in the statement: "If the system is in a proper state of its *observable A*, corresponding to the proper value *a*, then an *observer measuring A* on the system will *certainly obtain* the value *a*." The underlined words are out of place in a theoretical discourse for they point to the subject and some of his acts and mental states.

Moreover, such as it stands the above statement is false, because the typically quantum-mechanical properties are not directly observable (in the epistemological sense) and because measured values are usually only approximations to theoretically computed values. As to the certainty concept, it too is alien to a physical theory. A strictly physical theory, if

stochastic, must embody an objective, and particularly physical, interpretation of the probability calculus: it must interpret probability as a
physical property not as a measure of certainty (see Chapter 4, Section
2.4; Poincaré, 1912; Smoluchowski, 1918 and Popper, 1959). This does
not exclude the possibility of having psychological models of probability
theory: it just avoids, in the interest of consistency, mixing up the two
models. The postulate we have just criticised would have to be replaced
by something like this: "If the system is in a state represented by an
eigenstate of the operator representing its property A, then the numerical
value that A takes on is the eigenvalue a corresponding to that state."

Once the existing theory has been purged of all its nonphysical concepts,
it must be logically reorganised, if only to prevent relapses into subjectivism. There is no simple recipe to accomplish this task, for there are
several conceivable axiomatisations of any given axiomatisable theory.
(See, however, Chapter 6, Section 3.) The axiomatic foundation of QM
proposed by the author (Bunge, 1967a, 1967e) employs the following
primitive (undefined) concepts: "microsystem" (or *quanton*), "environment [microphysical or macrophysical] of the microsystem," "ordinary
(3-dimensional) space," "state space", "property of a microsystem," "operator representative of it" (the "observable" of the orthodox version), and
ten more which are much more specific concepts – among them those of
mass, charge, and energy operator. Every one of these concepts is then
characterised (not defined) by means of certain postulates – most of which
are far from self-evident and all are in the nature of hypotheses to be justified by the success of the theory in accounting for experimentally controllable facts.

The postulates of this realist version of QM characterise both the form
or mathematical nature of the basic concepts and their physical meaning:
the axiom system is consequently both formally and semantically determinate. Thus a conspicuous member of this axiom set states that certain
sets are nonempty and that their members are microsystems and their environments respectively. This physical truism is philosophically important:
it renders the theory nonvacuous and it commits it to epistemological
realism. Another axiom states that, if an operator represents a physical
property of a microsystem, then the proper values of that operator are
the sole values of the given property. Nothing is herein said about observations. Measurement will enter, as usual, at the test stage. For example

an aggregate of similar microsystems in a given milieu will be picked, some property of them will be measured, and the experimentally found frequency distribution (histogram) will be contrasted with the calculated probability distribution, concerning an individual system. Instead of dogmatically postulating that the experimental values are identical to the theoretical ones – as the Copenhagen doctrine does – the two sets of values will be compared. In case of discrepancy either the theory or the experiment or even both will be criticized.

Care will be taken throughout not to call either the microproperties or their conceptual representatives (the dynamical variables) *observables*. First, because they are not perceptible even though they are scrutable in an indirect way, much as impatience can be inferred from certain gestures and verbal utterances. Besides, to call the quantum-mechanical properties *observables* is to beg the important question concerning the design of means to measure them. Finally, as has already been suggested, the concept of observable is not a purely physical predicate, as shown by its analysis: "Object w is observable by subject x under the circumstances y with (empirical and theoretical) means z." If theoretical physics is not to be confused with psychology and epistemology, then the subject must be kept out of the former. The role of the subject is to build theories and test them, not to pose as their referent. For these reasons the dynamical variables occurring in QM should not be called 'observables'.

The typical quantum-mechanical magnitudes are random variables, in the sense that definite probability distributions are associated with them. This holds in particular for the position and the momentum of a microsystem – which should be called *quosition* and *quomentum* respectively to bring out their nonclassical character. That QM is basically probabilistic, is not assumed but proved in our version of the theory: in fact, it is shown that the function representing a quantum state satisfies the axioms of the probability calculus. In this way one sees that, in its present state, QM contains no hidden (i.e., nonrandom) variables. Thereupon von Neumann's celebrated proof of the impossibility of introducing hidden (scatter-free) variables in QM becomes a trivial metastatement obtained by merely scanning the primitive concepts of the axiom system and proving that all of those which function as dynamical variables are random variables. Any attempt to disprove von Neumann's thesis in connection with the current theory is then bound to fail just as miserably as any attempt to forbid the construction of alternative theories.

The basically stochastic character of QM can be understood in several ways. One is to assume that basic QM is not about an individual quanton but about a statistical ensemble of quantons: no wonder, then, that the various components of an ensemble in a given quantum state have different position and momentum values. Yet basic QM does work for the individual microsystem, e.g., for every one of the atoms crossing a crystal and impinging on a fluorescent screen. What happens is that the theory is tested by means of large assemblies of quantons; thus the computed position distribution is compared with the "diffraction" pattern that emerges on the screen as the number of individual impacts increases. In other words, just as any other random variable, the state function *refers* to an individual quanton (placed in a given milieu) but its precise form is *tested* with the help of statistical aggregates of quantons. And whenever aggregates of coexisting microsystems are concerned – especially if they interact – they must be handled by a more complex theory (a quantum statistics) based on elementary QM.

Another possibility is to assume that basic QM is neither about an arbitrary individual thing nor about an actual aggregate of similar things but about a conceptual set of such entities – a Gibbs ensemble (Bergmann, 1967). But this alternative does not seem to have been explored systematically. A third possibility is to regard quantum-mechanical properties as latent or potential rather than as actual, and as becoming actual or manifest upon the system's interaction with a measuring instrument (Margenau, 1950; Bohm, 1951). But this would render all dynamical variables dependent upon the observer, for they can become manifest at the observer's will.

However, the conception of quantum-mechanical properties as latent can be freed from its subjectivist tinge in the following way. As a rule, a quanton *has* neither a point-like position nor a sharp momentum; it has only precise position and momentum distributions. In general these distributions change in time under the action of the environment, whether or not the latter be under human control. In particular, a quanton can acquire a norrow spatial localisation. To this end, it is sufficient that manipulations aiming at the preparation of a localised state be made. But this is unnecessary: nature itself may do the trick once in a while – this being why we, a distinguished part of nature, sometimes succeed in spotting atoms or in producing nearly monokinetic electron beams. In all these cases a certain objective distribution becomes narrower, nearly point-like, and in this sense a classical property emerges or becomes actual – whereas its conjugate becomes classically less definite.

In a sense, this limit of the point-like distribution or classical property is a dispositional or potential property: indeed, the quanton has the possibility of acquiring it. But there is no potential/actual dichotomy in the Aristotelian style, for the distributions (in position, angular momentum, etc.) are properties possessed by the quanton all the time. Moreover, they are objective (subject-free) even though an observer may use real experimental arrangements to narrow down or to enlarge this or that distribution. All this involves, of course, forgetting about subjective probabilities and adopting one of the physical models of the probability calculus. In the quantum axiomatics proposed by the author, a modified (thoroughly physical) version of Popper's propensity interpretation is adopted. According to this theory, probability is a measure (not necessarily a measured value) of the objective disposition of a thing to behave in a certain way. If one wishes to avoid this interpretation, then he must work out QM as a theory about sets of replicas of an object, i.e. as a Gibbs statistics. But this has not yet been carried out. While this alternative is being explored, we may think of quantum dynamical variables as representing objective potentialities.

This must suffice to sketch the gist of our objectivistic axiomatic foundation of QM.

3. Seeing

A first advantage of this realist systematisation of QM is that it distinguishes the formal aspect from the semantical one, *i.e.* the quantum syntax from the quantum semantics. The physical content is poured into the theory via *interpretation hypotheses* – not just rules of designation but corrigible assumptions, and not "operational definitions" but objective and observer-free hypotheses. Like every other theory proper, QM contains theoretical concepts that fail to have an empirical interpretation, *i.e.* that cannot be introduced by means of "operational definitions." Moreover, none of the basic symbols of QM is empirically interpretable, whence the theory has no empirical content whatever: it describes no empirical item. This does not mean that QM is untestable:

it just means that the theory concerns transempirical facts rather than phenomena. In fact, the microfacts referred to by basic QM, such as quantum jumps, are imperceptible. The empirical tests of QM, like those of every other theory, require the assistance of further theories, of theories linking microfacts with macrofacts, and of theories accounting for the behaviour of the macrosystems (e.g., amplifiers) involved in measurements. In short, QM is *physically meaningful* because it refers to physical (though mostly imperceptible) entities and properties. And it becomes *empirically testable* if conjoined with specific subsidiary assumptions, experimental data, and further physical theories – otherwise it remains untestable. This would not have been accepted by those who used to mistake meaning for testability.

Our axiom system is both formally and semantically determinate. The interpretation of the formalism, as performed by the interpretive postulates of the theory, is, however, sketchy rather than full. Thus when stating that every state of the microsystem is represented by a point (or rather a ray) in a certain space (the Hilbert space of the system), the terms 'microsystem' and 'state' are neither defined nor described – other than by the postulates themselves. These words are taken from the physical jargon which the physical profession is supposed to master. Those and other words occur not only in QM but in other fields of physical science as well, and their meaning is specified jointly by all the fields of inquiry in which they occur. This is not peculiar to QM but common to all of factual science: here we lack the possibility, characteristic of mathematics, of interpreting one theory (e.g., group theory) in other theories (e.g., arithmetics and geometries).

The semantic or interpretive axioms of a physical theory pair off mathematical symbols to physical items – entities and properties thereof. Those of QM can be read analogically in terms of either particles or fields – or even fluids; and in some cases they can be interpreted in either way. This seems to have suggested to the fathers of the Copenhagen doctrine that there are two equally true and mutually complementary interpretations. In the author's view this only shows that the two interpretations are *ad hoc*: two different interpretations of one and the same formalism constitute two different theories, and two different theories can be compared to one another but they should not be mixed up. This contention is reinforced (a) by the nonexistence of a consistent axiomatisation of QM in either corpuscular or undulatory terms, and (b) by the fact that all the usual reasonings can be performed within our version of QM without ever employing the concepts of particle and wave. In particular, the state vector is not interpreted as describing a field strength. Neither is it interpreted as a field of knowledge. It is just a source of physical properties, much as the potentials and lagrangians. Upon dropping the classical analogies, QM undergoes a transformation similar to the one undergone by classical electromagnetism when special relativity showed that no mechanical ether was possible as a support for the electromagnetic field.

By dispensing with the classical concepts of particle (well-localised microbody) and wave (field ripple), we avoid the wave-field duality and the famed *complementarity "principle"*, a cornerstone of the Copenhagen doctrine. In our view the quanton is neither a classical particle nor a classical field but a *sui generis* entity that in certain extreme circumstances looks like a particle and on other occasions looks like a field. (See Section 2.) Whether these circumstances are natural or controlled by the experimenter is immaterial. In any event the concepts of particle and wave, legitimate though they are in connection with macrosystems (bodies and large scale fields), must be regarded as only metaphors at the quantum level – and, like every other metaphor, as double-edged tools: as heuristically valuable yet also misleading auxiliaries. The elimination of the wave-particle duality and the associated complementarity "principle" may be regarded as a further advantage of our formulation of QM, for in the name of complementarity too many inconsistencies and obscurities have been argued away. (More in Chapter 6.)

Another departed ghost is *uncertainty*. If QM is not about our mental states but – guess what? – about bits of matter and radiation, then the scatters occurring in Heisenberg's relations are not to be interpreted as subjective uncertainties but as objective latitudes in the quanton's localisation (Popper, 1959; Bunge, 1967a). Sure enough, our reformulation of QM does not eliminate uncertainty: one does not conquer infallibility just by axiomatising a field of knowledge. Only, the term 'uncertainty' is now pushed to some of the metalanguages of QM, i.e. it is allowed to occur in sentences concerning our skill in predicting facts with the help of QM: it does not occur in the object language of QM – nor should it occur in any other physical theory. *Certainty* is parallel.

The term *indeterminacy* to name the scatter around the average is slightly better than 'uncertainty' but not quite correct either, because there is nothing indeterminate in an objective position distribution as long as 'indeterminate' is equated with 'lawless and/or coming out of the blue' (Bunge, 1959a, 1962b, 1963). QM is stochastic all right and basically so, but a stochastic theory involving definite laws about probability distributions is not indeterministic if it makes no room for what is lawless and/or what is created out of nothing. In short, our version of QM is as deterministic as classical mechanics – only, it does not countenance laplacian determinism. Neither is the orthodox doctrine indeterministic: indeed, if quantum-mechanical probabilities are just degrees of certainty, then nothing can be inferred about the things in themselves. Ontological indeterminism requires a physical (objective) interpretation of probability. But as soon as the probabilities are both objective and lawful, indeterminism evaporates and stochastic determinism remains.

Something similar holds for other epistemological terms, such as 'observer', 'observable', and 'knowledge': they do not occur in the language of our theory although they may occur in any metalanguage of it – as when one says that the knowledge of the state a microsystem is in allows him to compute its momentum distribution and its average position. As a matter of fact this is the interpretation employed by the physicist except when he attempts to adapt QM to the official philosophy of physicists. Thus when he characterises the state vector he says that, for every microsystem in a given milieu, that symbol represents a function of space and time, and he adds that the form of the function may vary when the microsystem and/or its milieu change. In other words, both in our realistic interpretation of QM and in the daily work of the physicist, every point ψ in the space state is a complex valued function on $\Sigma \times \bar{\Sigma} \times E^3 \times T$, where '$\Sigma$' designates the set of quantons, '$\bar{\Sigma}$' the set of environments, 'E^3' ordinary space, and 'T' duration, the cross standing for the cartesian product. Something similar happens with the operators acting on ψ: in no case does the observer occur as an argument.

4. RESTORING OBJECTIVITY

The subject occurs, we said, in some of the *metalanguages* of the theory.

Take, e.g., a functional relation F between two variables x and y, every one of which represents a trait of a physical system σ of kind Σ. Since the traits related by F are physical properties of a σ, they must be represented by certain functions on Σ. Call them $g : \Sigma \to X$ and $h : \Sigma \to Y$, where X is the set of numerical values of g and Y the one of h. Then the functional relation F between x's and y's is some numerical function $F : X \to Y$ with values $y = F(x)$. Although the referents σ do not occur explicitly in the last expression, the latter must be read in terms of them because F is built out of f and g. Hence the formula "$y = F(x)$" should be interpreted as follows: the set X of numerical values of property g of system σ is mapped by F into the set Y of numerical values of property h of the same system. This is a strictly physical interpretation (schema) of the given functional relation. But the same formula may be reinterpreted in some metalanguage of the theory in which it occurs. For example, it may be reinterpreted in any of these two fashions: (1) Given F, for every x in X and every y in Y, *knowledge* of x uniquely determines y [in the epistemological not in the ontological sense of 'determination']. (2) For all x in X and all y in Y, y is *found* [or *computed*] from a suitable *measurement* of x by *use* of the formula: $y = F(x)$.

The last two interpretations may be called *epistemological* or *pragmatic*. The second of them is more restrictive than the first, which does not specify the kind of knowledge involved: this can be experimental or, indeed, it can cover the case of the hypothetical positing of values x. But either pragmatic interpretation is narrower than the physical one because it requires the presence of a cognitive subject – which, alas, is not everywhere and always available. The physical interpretation is the widest and moreover it is the basis or ground of the other two. First, because the epistemological interpretations belong to a metalanguage or the language in which '$y = F(x)$' occurs – and there is no metalanguage without a prior object language. Second, unless we are prepared to adopt solipsism all the way – i.e., to believe that the world is what we care it to be – we must assume that our knowledge is true to the extent that it models real things, relations, and events: if knowledge of x uniquely determines y via F, it must be because X and Y are in fact related in a unique way by F – i.e., because Y *is* the image of X under F, whether we happen to know this or not.

The ideal of *objectivity*, characteristic of factual science, is then shared by QM as much as by classical physics. The object has not disappeared, neither has it been welded to the subject. What has happened is that our current conceptual image of micro-objects is extremely sophisticated. And the items which are welded are not the object and the subject but the subject and his conceptual reconstruction of the object – but this has always been so except for Platonists. The subject does not occur among the basic predicates of our version of QM. Neither does he occur in the theory of measurement: indeed, physical theory is unconcerned with the psychical events going on inside the observer's skull: a physical theory of measurement is concerned only with the physical intersection between two or more physical entities, at least one of which must be a macrosystem.

True enough, on the usual version of QM the intervention of the observer produces a sudden contraction of the quantum state, which gets projected onto a proper state of the operator representing the "observable" that is being measured. Moreover, this collapse is allegedly lawless and therefore unpredictable, because there is no lawful relation between the original state and the final one. But this postulate leads to inconsistencies – to begin with, it is incompatible with Schrödinger's equation (von Neumann, 1932) – and is therefore not included in our formulation of QM. Besides, this projection postulate, asserting the collapse of the "wave function" at the sight of The Observer, implies the collapse of the principle of lawfulness, a basic ontological presupposition of scientific research (Bunge, 1967c).

The quantum theory of measurement ought to be built as an application of basic QM to the particular case in which the quanton is coupled to an unstable instrument capable of amplifying the microfacts that are to show up. Unfortunately, no such theory is available except in embryonic form, mainly because most physicists follow the mathematician von Neumann (von Neumann, 1932) in the belief that there are universal measuring devices, that is, instruments capable of measuring anything, so that their action can be represented by a single simple concept – the projection operator. But quite apart from technical considerations, the philosopher is competent to criticise the positivist thesis that QM is based on an analysis of the measurement process, as well as the more extreme claim that all of QM is about measurements. These theses are false for the following reasons: (a) no measurement can be planned and interpreted without the assistance of theories; (b) measurements involve macroprocesses, whereas basic QM concerns microevents; (c) for these reasons, the quantum theory of measurement, insofar as it exists, is an application of basic QM; (d) consequently any quantum mechanical statement concerning measurement must occur as a derivative statement, not as an axiom of the quantum theory.

A bonus of our version of QM is that it renders apparent the futility of the so-called quantum logic of Birkhoff and von Neumann (1936), Destouches-Février (1951), and others. The rationale for advocating this exotic logical calculus is this. If QM is true, then the propositions "The quanton x is at the point y at time t" and "The quanton x moves with velocity v at time t" are mutually incompatible, as shown by Heisenberg's latitude relations. It looks therefore as if QM embodies a logical calculus precluding the conjunction of certain ("noncommensurable") statements. But this argument stems from regarding quantons as classical particles. The difficulty does not occur if quantons are imagined as lacking, in general, both a sharp position and a sharp velocity but as having instead precise position and momentum distributions (see Section 2). This suffices to disperse the additional cloud of quantum logic. But it is not necessary: incompatible propositions occur everywhere, and odinary logic (the two-valued predicate calculus) is enough to handle such situations. If the conjunction of two propositions is false, all we have to do is to abstain from asserting it. Moreover, when axiomatising QM one starts by presupposing certain mathematical theories, such as analysis, which have classical logic built into them. Therefore, to accept classical logic at the foundations level, only to reject it at the level of theorems, is to fall into contradiction.

But it is high time to conclude.

5. Concluding

QM, one of the richest and deepest theories, has been befogged since its inception, forty years ago, by a subjectivist epistemology that goes back to Berkeley and Mach. This philosophical ballast is found not only in heaps of fashionable metastatements concerning QM but also in many of the very object statements of the standard version of the theory. As a result the referents of the standard version of QM became – to paraphrase Berkeley's indictment of Newton's shaky infinitesimals – the ghosts of departed physical entities.

It has often been claimed, with enviable conviction, that the marriage of QM to subjectivism, and particularly positivism, is indissoluble. This belief has led some to reject QM altogether, others to propose remodeling QM in a classical spirit, and most to live in the midst of the fog, either with fortitude or with joy. Meanwhile physicists have success-fully extended, applied and tested the basic theory, proceeding in their daily work without paying attention to the philosophical ballast. This mere fact should have suggested that the union of QM and subjectivist epistemology was a *mariage de convenance* whereby positivism enhanced its prestige while the new science, initially received with reluctance because of its departure from classical physics, enjoyed the support of a philosophy that was fashionable among scientists.

This marriage has now become a *mésalliance* and must be disolved. Indeed, (a) the subjectivist epistemology espoused by logical positivism is now dead or nearly so as a consequence of both external criticism and the honest self-criticism performed by the positivists themselves; (b) one can eliminate the subjectivist ballast encumbering QM, turning the latter into a thoroughly physical theory – a theory free from psycholog-ical elements. By so doing QM has not remained a bachelor but has taken a new philosophical spouse: realism. Not of course uncritical realism, but a realism that, while postulating the autonomous existence of the external world, is ready to correct every conceptual reconstruction of it – a realism acknowledging that even though we intend to map sections of reality, we do it in a piecemeal, imperfect and symbolic way rather than completely and literally. (Recall Chapter 4, Section 4.)

This, far from producing complacency in the realist camp, should set it in motion: while QM is no longer living proof that realism is untenable, it does suggest that the existing varieties of realism are under-developed because they fail to offer a detailed description and analysis of the sophisticated ways in which scientific research proceeds to invent and test conceptual models of chunks of reality. The metaphysician should feel a similar challenge. Up to now he had been told that QM proves matter to be more like mind than like matter, whereupon he was either baffled or delighted. He should now realise that matter has not been dematerialised by QM (Feigl, 1962) but that the picture physics draws of it is far more complex than what mechanics and classical field theory had assumed: quantons are proteic things than can hardly be pictured in classical terms. But in any case they are out there, at the door of ontology, demanding a new look at certain basic ontological categories such as those of substance, form, motion, novelty, determination, causation, chance, and law. May the new physics, once cleansed of an obsolete philosophy, stimulate new developments in epistemology and ontology. (For suggestions in this direction see Bunge, 1972a.) And may the new philosophy help scientific progress rather than hinder it. One contribution it can already make is to help separate the theoretical grain from the heuristic chaff – as will be shown in the next chapter.

CHAPTER 6

ANALOGY AND COMPLEMENTARITY*

Analogy is undoubtedly as prolific as it is treacherous. It can serve essentially three useful purposes. *Heuristic use*: for classing, generalising, finding new laws, building new theories, and interpreting new formulas. *Computational use*: for solving computation problems by handling analogues (e.g., electrical models of mechanical systems). *Experimental use*: for solving problems of empirical test by handling analogues, particularly replicas and simulates (e.g., experimental stress analysis of steel bodies on translucent plastic replicas). We shall focus here on the first function of analogy, as found in the quantum theory.

The assertion that the state vector of a microsystem represents a real wave (the original de Broglie–Schrödinger interpretation) was jumped at from the formal analogy between the quantum-mechanical state equation and the classical wave equations: it was an analogical inference based on an analogy of the formal kind. (For elucidations of the concepts of formal analogy and substantial analogy see Metzger, 1926; Bunge, 1967c, 1967f, and 1972a.) In the early days of the theory this interpretation was taken as literal not metaphorical, i.e., the formal analogy was taken as indicative of a substantial analogy. Shortly thereafter, by proposing his stochastic interpretation, Born tacitly showed that no substantial similarity was to be concluded from the formal analogy: that the state vector does not represent a peculiar substance spread over the whole space available to the system but that it represents the state of the system. Granted, without the analogy wave mechanics might not have been born and the so-called diffraction of matter waves would not have counted as a decisive empirical confirmation of it. But the survival of this interpretation alongside the stochastic interpretation (in turn cast in a corpuscular language) is responsible for many of the conceptual (noncomputational and nonempirical) difficulties of the quantum theory. This will be shown in the sequel.

1. Analogy double-edged

There is no doubt that analogy can be fruitful in the preliminary exploration of new scientific territory, by suggesting that the new and unknown is, in some respects, like the old and known. If B behaves like A in certain regards, then it is worthwhile to hypothesise that it does so in other respects as well. Whether or not the hypothesis succeeds, we shall have learned something, while nothing will have been learned if no hypothesis at all had been formulated. Should an analogy hypothesis pass the tests, we shall know that A and B are indeed similar either substantially or formally. And if the analogy fails utterly, we shall realise that some radically new ideas are called for because B is in some respect radically different from A. But unless the analogy is extremely specific or detailed, chances are that it will hold to a first approximation for, after all, our conceptual outfit is limited and no two concrete systems are dissimilar in every respect. The question is to decide what to stress at a given stage of research, i.e., whether resemblance or difference should be emphasised.

The elementary treatment of the scattering of light by electrons (Compton effect) is a typical example of the initial triumph and final failure of analogy as a method for handling newcomers. By feigning that both the electron and the photon are particles (an assumption absent from the advanced treatment), the problem can be reduced to a two body elastic collision and the formula for the frequency shift of the scattered radiation can be obtained – as long as one is willing to put up with a frequency that is incomprehensible in the context of mechanics. That is, if we restrict our attention to momentum exchanges and neglect all the other questions, then the photon-ball analogy – which, though baffling, is both substantial and formal – proves fertile. But it breaks down as soon as further questions are asked, some of which will receive wrong answers suggested by the analogy while others will receive no answer at all for they are not covered by the analogy. Thus the analogy will suggest asking what the photon mass is, the wrong answer being that it must be $h v/c^2$ since its momentum equals $h v/c$ and its velocity is c. Which is nonsense, for there is no relative mass without a rest mass, and no rest mass without a rest frame – not to mention the nonexistence of photon equations of motion. On the other hand the mechanical analogy will not help us compute the scattering cross section, which requires the consideration of some of the electromagnetic and quantum-mechanical properties of the electron and the photon.

Analogy, then, is double-edged. On the one hand it facilitates research into the unknown by encouraging us to tentatively extend our antecedent knowledge to a new field. On the other hand, if the world is variegated then analogy is bound to exhibit its limitation at some point, for what is radically new is precisely that which cannot be fully accounted for in familiar terms. This seems to be the case with the analogies that helped build the quantum theories, particularly the particle and the wave analogies: they seem to have reached their breaking point long ago. No doubt, in the beginning it was just human to seek inspiration in classical physics, making the pretence that electrons and photons behave sometimes as particles and at other times as fields: there was no help other than analogy for assigning wave mechanics and matrix mechanics (and later on quantum electrodynamics as well) some physical meaning. But in the meantime we should have learned at least two lessons. The first is, that the particle and the wave analogies are weak and moreover mutually inconsistent. The second, that the entities the quantum theories are about behave in an original way, i.e., according to nonclassical laws – whence they can be neither classical bodies nor classical fields. Therefore it is high time we acknowledged that the quantum theories should get rid of those classical analogues and should recognise that they are concerned with *sui generis* things deserving a new generic name – say, *quantons* (Bunge, 1967a).

2. The Wave-particle Duality in Optics

Several wave theories of light reigned in optics from the time of Young, Fresnel and Cauchy to the birth of the photon hypothesis in 1905. From here on until the creation of quantum electrodynamics in 1927, two mutually inconsistent bodies of ideas were used to account for the facts of optics: Maxwell's field theory, and a set of hypotheses (hardly a hypothetico-deductive system) clustering around the photon hypothesis. The nature of light was thought to be dual and this duality was often regarded as irreducible.

Quantum electrodynamics (henceforth QED) was built with the

purpose of incorporating the wave-particle duality into a consistent body of ideas, thus overcoming the manicheism of the previous period. It is still widely believed that QED did succeed in incorporating that duality, mainly because the theory assigns the photon both a linear and an angular momentum (or rather quantal analogues of such). But of course this does not establish that QED regards the photon as a particle. First, because QED denies it the two defining corpuscular properties, namely a sharp localisation – hence a definite path – and a mass. Second because, as a consequence, QED contains no equation of motion proper: all of its basic equations, including the commutation relations, are field equations, none of which entails photon trajectories in ordinary space. (True, any formula for the time rate of change of a dynamical variable is called an equation of motion, but this is just a metaphor, for there need be no motion involved.) Third because the properties QED assigns the radiation field are nonmechanical – e.g., the electric and magnetic components, the phases, and the frame-independence of the field propagation speed.

If anything, QED has remained closer to Maxwell's field theory than to mechanics: after all, it is essentially the quantisation of the former. The quantum properties of the field should not be mistaken for mechanical or corpuscular properties of it. Thus the fact that the photon momentum can be added to the mechanical momentum of a chunk of matter, to yield a conserved quantity, does not prove the mechanical nature of photons any more than it proves the electromagnetic nature of matter; likewise the possibility of adding energies of different kinds does not prove their identity with mechanical work. All this shows is that the momentum-energy four-vector, unlike mass and charge, is a nonspecific property, i.e., one characterising every physical system known heretofore. Likewise the fact that the field energy can be decomposed into the energy of radiation oscillators does not prove that the field is a mechanical system but rather that the hamiltonian formalism is noncommittal and lends itself to mechanical analogies that are sometimes misleading (Bunge, 1957). If we cast different theories into one and the same mathematical framework – say hamiltonian 'dynamics' – then we are bound to produce formal analogies. But then we should not hypostatise this mathematical similarity by exclaiming: Lo and behold, mother Nature is all mechanical (or all electromagnetic)!

In short, the wave-particle duality that had broken the unity of electro-magnetic theory in 1905 stimulated the creation of another theory, QED, that eliminated that duality. Even in its most sophisticated versions QED is a field theory containing no hypothesis about the corpuscular nature of photons. The optical duality is then a relic of the 1905–1927 interregnum, a remains serving mainly to mislead students into believing that light is at the same time undulatory and non-undulatory or, even worse, that it is so devoid of character that it will look what the omnipotent observer chooses to.

3. The Wave-Particle Duality in Quantum Mechanics

The wave-particle duality suggested to de Broglie the problem whether matter might not exhibit a similar duality. If the electromagnetic field seemed to have a mechanical side, it was just possible that matter had a field-like aspect as well. "The idea of such a symmetry was the point of departure of wave mechanics" (de Broglie, 1937, p. 171. See also Schrödinger, 1926, p. 489). That the analogy was extraordinarily fruitful does not prove that it was substantial rather than formal.

It was fortunate that de Broglie and Schrödinger did not seem to realise at the time that the formalisms of Hamilton and Hamilton-Jacobi are generous enough to harbour almost any physical theory, from mechanics to thermodynamics. Had they known this they might not have marvelled at the analogy between Fermat's optical principle and Hamilton's mechanical principle and they might consequently have missed inventing wave mechanics. We would have had to wrestle with the less pictorial matrix mechanics and we would have been spared the mistake of believing that quantum mechanics deals with waves of a certain kind (matter waves).

We have come to realise the originality of quantum mechanics (henceforth QM) because the theory can be formulated without resorting to the heuristics of de Broglie and Schrödinger. Moreover, although when working in this particular formulation (the Schrödinger "picture") one still uses expressions like 'wave function', 'wave packet', 'wave length', and 'wave equation', one tends to regard these as one-sided classical analogues. The very phrase 'the de Broglie wave associated to an electron' shows that we no longer believe electrons to *be* waves: we tend instead to think of the state vector as a global and basic property of a physical system rather than

wavelength is Galilei invariant while the momentum p is not. And the dualistic interpretation of the second formula is inconsistent with Born's stochastic interpretation of the state vector, which entails that 'Δx' (and likewise 'Δp') designates the standard deviation or statistical scatter around the mean value, a stochastic concept that has little if anything to do with wave packet sizes, apertures of slits in diffraction arrangements, and other items of the dualistic interpretation (Landé 1965; Bunge, 1967a).

Yet the standard textbook will make no effort to tally the preceding formulas with Born's stochastic interpretation, which they contradict. In particular, it will waver between these mutually incompatible interpretations of 'Δx' and 'Δp': the objective position (momentum) indeterminacy of the *particle*, the spatial (spectral) width of the *wave* packet associated to the *particle*, the size of the disturbance caused by the apparatus on the *particle* position (momentum), and our objective uncertainty concerning the actual position (momentum) of the *particle* diffracted (*wave*-wise) through a slit. It is seldom realised that this erratic behaviour is inconsistent and therefore unscientific. Nor is it usually pointed out that every one of these interpretations is *ad hoc*, in the sense that it is arbitrarily forced upon the symbols concerned without any warrant for it. Indeed, firstly, the postulates of QM do not state that quantons are particles – nor do they state that they are field chunks. And secondly, no assumptions concerning either an apparatus or an observer are ever made in order to derive the Heisenberg relations from the postulates of QM, whence it is logically impermissible to mention them at all at the level of the general theory: to do it is to incur in a semantical inconsistency. (Recall Chapter 5, Section 1.)

At other times it is stated that QM consists of two mutually equivalent theories, one of them cast in the particle language, the other in the wave language. This too is a mistake. There are many, not just two, different formulations of QM and most of them are mathematically isomorphic – wich does not entail that they are assigned the same meaning. One of these formulations is Schrödinger's "picture", in terms of a time-dependent state vector; another is Heisenberg's "picture", in terms of time-dependent dynamical variables. Further formulations are the density matrix and the path integral ones. They are equivalent formulations of one and the same theory and they are not pictures but high-level symbolic constructions. The Schrödinger formulation suggests analogies with classical field theories; the Heisenberg and the Feynman formulations invite analogies with classical particle mechanics; and the density matrix formulation suggests, if at all, analogies with classical statistical mechanics. *Pace* Dirac (1972) there is no best "picture" (Heisenberg's) for "understanding" QM. And none of these interpretations can be carried through consistently: these analogies are formal, they concern form similarities among some, not all, the quantum-mechanical formulas, and certain classical formulas. Moreover, these analogies cannot be carried over to QED.

In short the optico-mechanical analogy, which used to be a fertile working hypothesis, should now be abandoned for it has outlived its purpose by becoming a source of confusion.

5. RISE AND FALL OF COMPLEMENTARITY

In the mid-1920's the most advanced physicists believed they had to put up with two dualities: the alleged dual nature of the electromagnetic field and the possible duality of matter. From this double duality there was but a short step to the sweeping ontological conjecture that every physical entity has both corpuscular and undulatory aspects. This is the thesis of *general dualism*; it is a metaphysical hypothesis for it concerns the basic nature of every existent. When QM was built, Heisenberg's scatter ("uncertainty") relations were interpreted in the light of dualism: they were made into a precise and striking illustration of dualism. Bohr's *complementarity principle* – which, like Mach's principle and the Monroe doctrine, was never stated unambiguously, let alone understood clearly – was a specification as well as a reinterpretation of general dualism. It was a specification or particularisation for, in addition to stating the duality, it asserted that the more one of the two aspects is enhanced the more its complement is blurred: the more *yin* the less *yang* and conversely.

But, unlike the ontological thesis of general dualism, the complementarity principle claimed to concern the subject-object complex rather than autonomously existing microsystems. Indeed, the orthodox formulations of the principle do not assert that the corpuscular and the undulatory traits of a microsystem balance one another. Instead, they state that what can be complementary is either a pair of macroscopic experimental arrangements (the observer included) or a pair of descriptions of the outcomes of operations conducted with the help of such laboratory set-ups or, finally, a pair of concepts. Complementarity, in short, strengthens dualism by rendering it slightly more precise – though not precise enough – but on the other hand it weakens duality by failing to attribute it to nature: things in themselves, e.g. atoms in free space, would not have a dual nature – moreover they would be just figments of an imagination not disciplined by the subject-centered philosophy of Copenhagen.

Moreover, since the experimental arrangements and their results are supposed to be classically describable, the complementarity principle remains on this side of QM and QED: strictly speaking it is not a quantum-theoretical statement insofar as it does not concern microsystems. If regarded as a principle of either QM or QED, then it comes in contradiction with the assertion that microsystems satisfy nonclassical laws and must therefore be described and explained in nonclassical terms. But strictly speaking the complementarity principle is not a principle, for it entails nothing. In fact, no theorem follows from it. In order to prove a theorem in quantum theory one processes a bunch of axioms usually in conjunction with a set of specific hypotheses concerning, say, the number of microsystems and their mutual actions – in short, one adds a definite model to the general assumptions: see Section 8 – but one does not use the complementarity principle, which is much too sweeping and vague to entail anything. (In particular, this pseudoprinciple finds no application in QED, for the statistical spreads of the components of the electromagnetic field defy a corpuscular interpretation.) The complementarity pseudoprinciple is then neither a principle nor a theorem – nor is it as general as usually claimed, for it does not hold for fields. And in the advanced quantum theory of "particles" (second quantisation) the field is treated as the primordial thing. Thus in the case of electrons, or of mesons, the matter field is regarded as the primary entity whereas the "particles", or rather the particle-like entities, are just field quanta, i.e. field chunks. (An eigenvalue of the occupation number operator represents the number of entities in a given state, and these entities – the field quanta – are not classical particles.) In other words, any second-quantised theory is closer to a classical field theory than to classical mechanics even though it can be cast in a hamiltonian or a lagrangian framework. Consequently there is no place for complementarity in the more sophisticated areas of the quantum theory. Nor is there, *a fortiori*, in any of the phenomenological theories – such as the scattering matrix formalism – that eschew a detailed description of the field.

What then, apart from authority, explains the survival of complementarity? The main reason seems to be its great usefulness. Indeed, complementarity explains many difficulties away and it accounts for experiments of two kinds: gedanken-experiments that have never been carried out and real experiments that have never been computed in quantum-theoretical terms. The first point is obvious: once the pseudo-principle is accepted, it can be used to consecrate obscurities and inconsistencies, much as the mystery of the trinity subsumes many minor mysteries. As to the experiment that allegedly illustrate the principle, they are in fact either imaginary or still beyond the scope of the theory. Among the former one finds Heisenberg's gamma ray microscope and Bohr's shutter experiment in his discussion with Einstein. Since they have no confirming power we may leave them aside. Among the experiments of the second kind, the diffraction experiments stand out. Unfortunately the diffraction by a single slit has been calculated only for an infinitely long slit and a monochromatic de Broglie "wave." Moreover the available computation is approximate and its result conflicts head-on with the Heisenberg inequalities (Beck and Nussenzveig, 1958) – which of course cannot be held against the latter. *A fortiori* the much-discussed double slit experiment has never been computed exactly in QM, let alone in QED. So much so that the most complete, accurate and recent 681 pages long work on scattering (Newton, 1966) does not treat the standard diffraction gedanken-experiment found in discussions on the quantum theory. (On the other hand, and oddly enough, they can be explained in a purely corpuscular fashion, with the help of Bohr's classical quantum theory, as an effect of the periodicity of the scattering lattice (Landé, 1965).) True, a few formulas are used in the qualitative discussions of these experiments, but they are taken from the general theory: they do not result from an application of it to those special circumstances. Also, some nice diffraction patterns are shown, but they are taken from real (but so far uncalculated) experiments or they are borrowed from classical optics. In short, the discussions of these experiments in terms of complementarity is verbal and analogical. Consequently complementarity is not a part and parcel of the quantum theory.

In conclusion complementarity, reasonable though it looked in the early days of the quantum theory, when people thought in terms of classical pictures, has now exhausted whatever potency it may have had: it has become an excuse for obscurity and inconsistency.

6. TOWARDS A LITERAL INTERPRETATION OF THE QUANTUM THEORY

A number of physicists have felt that the concepts of wave and particle do not belong in QM because they are classical metaphors. Thus Schrödinger held at one time that wave and particle are "images we are forced to keep because we do not know how to get rid of" (Schrödinger, 1933, p. xiv). The impression that these are just metaphors, images or visual props, is reinforced by the fact that they fail to occur in any careful attempt to formulate QM in an orderly fashion, i.e. axiomatically. Indeed, they occur neither as undefined concepts nor as derived ones; hence they should not be allowed to figure in the theorems either. The main reason those classical analogues still play an important role in foundations discussions of QM and even QED seem to be inertia and partisan spirit.

Although most of us realise that the quantum theories chart new territory, we persist in trying to understand it in classical terms – much as Columbus called Indies the lands he discovered because he did not realise how novel his discovery was. We find it handy to call ψ a *wave* function and to draw pictures of wave fronts – only to point out that the "wave" is a complex function and that its main duty is to inform us where the *particle* is likely to be found. We find it intuitive to call the Schrödinger equation a *wave* equation – only to add that it governs the propagation of ψ in a $3N$-dimensional space. We find it convenient to speak of the *diffraction* of *particles* by a crystal (why not of the collision of waves?) and of the phase shift of the *particle's* associated *wave* produced by an external field. We thus keep piling inconsistencies hoping that the complementarity principle, like confession, will absolve us.

But we can do even better: with hindsight and ingenuity classical physics can be restated in ways reminiscent of the quantum theory. Thus particle mechanics can be formulated, within the Hamilton-Jacobi framework, as a field theory dealing with the propagation of a fictitious wave built with the help of a solution of the equation of motion. If classical analogs of second quantisation is required, they can be supplied as well (Schiller, 1967; Bourret, 1967). In short, just as nearly every nonrelativistic quantum-mechanical formula can be misinterpreted in classical terms, so nearly every classical formula can be restated in a (pseudo) quantal fashion. Unfortunately, at this late hour few of these analogies are more than formal games: they seldom yield new insights and they never lead to new specific predictions. Trying to discover the quantum in classical physics is as hopeless as trying to read QM and QED in classical terms.

The evasion from this tangle of inconsistencies, obscurities and metaphors is simple enough, namely to regard microsystems as altogether heathen individuals. Consequently they should be rechristened with heathen names, such as *quanton* (family name) and the genera names *hylon* (from ὕλη, matter) and *pedion* (from πεδίον, field). Even the names of the theories concerned could be profitably changed, for instance into *hylonics* (=quantum mechanics), *pedionics* (=quantum field theory) and *quantics* (the union of the two). After all, the quantum theory is a successful upstart and every successful upstart needs a new name that will not betray its origin.

Of course this is not just a question of names: classical concepts must be either reformed or wiped out from the quantum theory unless they happen to function in it just as they did in classical physics. Thus in the elementary theory the concepts of mass, charge and electromagnetic field are classical. On the other hand what is usually called the operator representative of the particle position is nothing of the sort: 'x' designates just a point in the configuration space and, unless the probability distribution is given, a specific value of x tells us nothing about the hazy location of the quanton. Only the quantum-mechanical average of x, built with the help of that probability density, is the analogue of the classical position coordinate, as shown by both the formal structure of the concept and by the formal analogy between the corresponding equations of motion. The foundations of QM and QED can and must be laid down without the help of either classical analogies or of more or less ideal measurements – just as thermodynamics is formulated nowadays without the caloric fiction and without resorting to engine cycles.

It is only when looking for classical or semiclassical correspondence limits, and when applying the general theory to particular cases, that we are entitled to seek again inspiration in classical physics. We may try to find quantum renditions of classical formulas and, conversely, classical analogues of quantum formulas. (An expression C may be called a *classical analogue* of a quantum-theoretical expression Q if and only if either C and Q play homologous roles in formally analogous formulas, or C is a correspondence limit of Q.) And we shall also avail ourselves of classical physics when hypothesising quantum hamiltonians or lagrangians. Borrowing them from classical physics and rewriting them in quantum-mechanical terms with the help of heuristic rules is honest practice provided the new hamiltonians are not read in classical terms. (However, stealing does not pay in two important cases: when the classical hamiltonian cannot be symmetrised in an unambiguous way, and when radically new interactions, such as exchange forces, are at stake.) In any case, whether or not a quantum formula has a classical analogue, it should not be interpreted in classical terms but as dictated by the interpretation axioms of the theory. And these assumptions (also called 'correspondence rules' and sometimes 'operational definitions') should be literal rather than metaphorical, as well as objective rather than operator-centered.

A physical theory is assigned a *literal and objective interpretation* by assigning every one of its referential primitive symbols a physical object – entity, property, relation, or event – rather than a mental picture or a human operation. Thus the state vector is not properly interpreted as a wave field (in the style of the electromagnetic field) or as an information carrier, but as representing the state of the system concerned – just as in statistical mechanics every state of a N-body system is mapped on to a point in the corresponding $6N$-dimensional phase space. The fact that the evolution of the state of the system is described (in the Schrödinger formulation) by an equation reminding us of a wave equation, does not prove the existence of a substantial analogy, the more so since many alternative metaphorical interpretations are conceivable (Bunge, 1956).

As with foundations so with their applications: analogical thinking, however fertile as a starter, eventually becomes confusing. A case in point is many-body theory with its twenty or so quasiparticles and pseudoparticles. Thus on an analogy with the electromagnetic field it was conjectured that sound waves are quantised, i.e. that the kinetic elastic energy of a body equals an integral number of sound quanta or phonons. This hypothesis was eventually expanded into a full-blown theory which found abundant empirical confirmation. But the analogy, though fruitful, is shallow: while a photon is a chunk of electromagnetic field and may acquire an autonomous existence by freeing itself from its source, a phonon is no such independent thing: it is a property of a complex system – there are no free phonons. Similarly with the other quasiparticles and with the so-called resonances in elementary particle theory: they are states of things rather than independent things. To suggest that they behave *like*, or *as if* they were, particles was insightful, for it started the conceptual machinery. To maintain that they *are* particles is nonsense, for they fail to exhibit the defining characteristics of a particle, namely independent existence, a sharp localisation, and a mass. Yet the current physical literature teems with such analogical nonsense.

7. Rigorous interpretation and explanation – literal not metaphorical

Poets, theologians and occult scienticists resort to metaphors and analogies on the ground that they handle subjects defying direct description or perhaps even rational understanding. Teachers use the same trick for a different reason, namely in order to bridge the chasm between the unknown and the familiar. Who among us has resisted the temptation to picture electrons sometimes as balls and at other times as wave packets? Yet we know that such metaphors and analogies are at best didactic props, at worst didactic traps, always *Ersätze* for the real thing. We therefore try to avoid them in research. We wish science to be concerned with what things are not with what things look like: science is neither poetry nor theodicy nor occult science. While we are willing to let analogy guide our preliminary explorations (notice the metaphor), we feel it is wrong to allow it to play any role in a mature theory: we want this to depict the thing itself rather than something that superficially looks like it. In other words, we want *literal interpretations* – even if they are assigned no familiar visualisations – because we want objectivity. It is only in mathematics that we are interested in mirroring one conceptual structure on to another. The conceptual frameworks of factual science are supposed to mirror (symbolically and partially to be sure) real things not further constructs. To stick to analogy in factual science is just beating around the bush: analogical thinking is characteristic of proto-science (e.g. history) and pseudoscience (e.g. psychoanalysis). Grown-up science is literal as much as it is objective. Its epistemology is consequently realist. (See Bunge, 1972a, Chapter 10.)

A *literal and objective interpretation* of a basic (primitive) symbol s occurring in a physical theory T assigns s a physical object p, be it an entity (e.g. an atom), a property thereof (e.g. the atom angular momentum) or a change of it (e.g. a jump in the angular momentum value). In short, $p = Int(s)$. A literal and objective interpretation of a whole physical theory T will accordingly consist in a mapping $Int: S \rightarrow P$ of the set S of the basic symbols of T into the set P of their physical partners. If P lies in another field, covered by a theory T' (i.e. if $P \subset P'$, where P' is the set of physical objects referred to by T'), then T will be interpreted in *analogy* with T'. In particular, if T is a quantum theory and T' a classical theory and P is included in the referent of T', then the classical interpretation of T will be metaphorical. And if P has a nonempty overlap with a set of psychological objects such as human dispositions and abilities, say observability, uncertainty, and predictability, then T will be a psychophysical rather than a strictly physical theory.

What holds for interpretation holds for explanation: if the formalism of a physical theory is assigned a literal and objective interpretation, then any explanation contrived with the help of that theory will be literal as well. We will not reject metaphorical explanation altogether: it should be tolerated, *faute de mieux*, in the period of theory construction. Thus it would have been foolish to reject the hydrodynamical analogies of heat and electricity "flow" in the beginning – as foolish as regarding them as substantial rather than formal. Likewise information theorists did well to profit by the formal similarity between information and negentropy. But it is abusive to reverse the move and reduce entropy increase to loss of human information about the system, for this deprives statistical mechanics and thermodynamics of objectivity. In other words, an interpretation of 'S' in statistical mechanics must assign it an objective physical property not the state of human knowledge. Likewise, if QM and QED are to count as physical theories then ψ must be assigned a physical meaning in an objective and literal fashion.

Quantum theoretical explanations, when stripped of supernumerary

observers and measuring instruments, are certainly less intuitive than
many a classical explanation came to be after having been regarded as
unclear by Leibniz and the Cartesians with their insistence on explaining
everything in terms of shape and motion. Yet quantum theories explain
in a perfectly orthodox sense of 'explanation', namely as deduction from
general assumptions (in particular laws) and particular premises (e.g.
data). But, just as thermodynamics, continuum mechanics and electromagnetic theory cannot be understood in corpuscular terms and are
quite unintuitive and occasionally even counterintuitive, so the quantum
theories demand new ways of understanding. Bohr went so far as to
assert that a new concept of understanding is at stake. He would have
said that "Through quantum mechanics the meaning of the word
'understanding' has changed in such a way that here, too, we have to do
with a complete understanding of quantum phenomena" (Heisenberg,
personal communication, 1970). I would put it this other way: We have
achieved a satisfactory *explanation* of many (not all) quantum phenomena
even though we are unable and even unwilling to *understand* them in
traditional terms, particularly in terms of classical concepts like those
of particle and wave, which do not belong by right to quantum mechanics.
Anyhow understanding is a psychological not an epistemological category.
And those who are disappointed by the little "understanding" quantum
mechanics brings them should feel soothed upon being told (a) that the
more they stick to heuristic pictures the less they will understand the
theory and (b) that they should not expect to find a kinematics, or
theory of motion, in quantum mechanics just because one gives it the
misnomer *mechanics* (recall Section 6).

The situation, then, is this. If we want to build or learn new theories
then we are likely to use analogy as a bridge between the known and the
unknown. But as soon as the new theory is on hand it should be subjected
to a critical examination with a view to dismounting its heuristic scaffolding and reconstructing the system in a literal way – this being one
of the uses of axiomatisation. Once this reconstruction has been achieved,
i.e. once we have found out not what the theory looks like but what it is,
and not what the referents of theory mimick but what they are, we must
refuse to acknowledge any metaphorical explanation in the field covered
by the theory, for it will be a pseudoexplanation. To suggest that scientific
explanation is metaphorical is to mistake scientific theories for biblical

parables or to subscribe to instrumentalism, for which "All knowledge, unless it establishes just actual succession and coexistence, can only be analogical" (Vaihinger, 1920, p. 42).

8. MODELS

How about models: should they too be regarded as mere heuristic auxiliaries to be discarded after the theory has been built? The answer depends on the sense of the polymorphous word 'model', a term which is as widely used as little analysed in the recent philosophy of physics. There are two senses in which models are indeed an ingredient of physical theories and two others in which they need not be so.

If 'model' is taken to mean *visual representation* or analogy with familiar experience (Hutten, 1956), then clearly not every theory involves a model. Thus field theories, whether classical or quantal, are hardly visualisable. And if 'model' is taken to mean *mechanism* – either in a narrow mechanical sense or in a wide sense including nonmechanical mechanisms such as the meson field mechanism of nuclear forces – then some theories do contain models of this kind while others do not. (The former may be called mechanism or representational theories while the latter may be called phenomenological or black box theories: d'Abro, 1939, Bunge, 1964b.) Consequently the neo-Kelvinist view that every scientific theory contains or presupposes a model consisting in a pictorial representation or analogy, is not adequate. Therefore the associated view, that both scientific interpretation and scientific explanation require visual representations, is false as well.

On the other hand in a third sense every physical theory *is* a model, namely of the underlying mathematical formalism. Moveorer a physical theory is twice a model in the model-theoretic sense: once because every one of its basic signs has a particular interpretation within mathematics, another time because the same sign may have a physical interpretation as well – as is the case with all the referential primitives. Thus in mechanics 'm' may first be interpreted as a number, then as the mass of the body it is assigned to. The final interpretation of 'm' is then the composition of a mathematical and a physical interpretation function: the sign is assigned a number which is in turn interpreted as a mass value. *Mutatis mutandis* for the whole symbolism of the theory. Caution: this, the model-theoretic sense of 'model', can be used only in connection with axiomatised theories, for a model of an otherwise uninterpreted factual theory is built by assigning every one of the primitives of the theory a factual interpretation – and one does not know which are the primitives before axiomatising the theory. But because the interpretation is factual and the resulting interpreted formalism is satisfied only approximately (if at all) by its referents, the model theoretic concept of a model is of little interest in relation to theories in factual science. (See Bunge, 1972a.)

And in a fourth sense every *specific* physical theory (but not every theory) *contains* a model or sketch of its particular referent. (Recall Chapter 3, Section 4.) The general formulas of a theory are nonspecific, so that they do not suffice to solve particular problems, such as finding the trajectory of a missile or the modes of propagation of a wave in a waveguide or the energy levels of an atom. In order to solve any such specific problem a number of specific assumptions and data concerning the particular physical system referred to must be supplied: the number and nature of the parts and their assumed interactions, the constraints and the constitutive equations, the initial and boundary conditions, and what have you. And these subsidiary hypotheses and data, which are adjoined to the generic axioms of the theory to cure their indeterminateness, jointly constitute a *conceptual model* of the concrete system. A model in this fourth sense is then a set of *statements* specifying (roughly) the nature of the referent of the theory in a more precise way than the general (and therefore highly indeterminate) assumptions do.

Here go some examples of conceptual models in physics: (1) the hard-sphere model of a gas; (2) the Ising model of a body in a condensed phase: an array of atoms or molecules regularly spaced such that every one of them interacts only with its nearest neighbours; (3) the classical model of a fluid, or even the whole universe, as a continuous medium with given density and stress destributions; (4) the elementary model of an electric current as a one-dimensional current of infinite density; (5) the potential barrier as a representative of an external force and the potential well as a schematisation of an attractive inner force in quantum mechanics. Notice, first, that every such model involves some of the concepts of a given theory, for otherwise it could not be adjoined to it. Second, none of these models is committed to a specific set of law statements: they can occur in widely different and even mutually incompatible theories of a given class (classical and quantal, nonrelativistic and relativistic, etc.). In short, the conceptual model is not part and parcel of the foundations of a general theory: this theory must be on hand, as fully interpreted as possible, if it is going to be applied to a model thereby becoming a specific theory – e.g. a nonrelativistic quantum theory of the helium atom. That is, the model does not contribute, except heuristically, to endowing the general theory with a factual (e.g. physical) meaning. Third, a conceptual model is neither idle nor faithful: it is, or rather it is supposed to be and so taken until further notice, an approximate representation of a real thing.

Whether or not a given conceptual model or representation of a physical system happens to be picturable, is irrelevant to the semantics of the theory to which it eventually becomes attached. Picturability is a fortunate psychological occurrence, not a scientific necessity. Few of the models that pass for visual representations are picturable anyhow. For one thing, the model may be and usually is constituted by imperceptible items such as unextended particles and invisible fields. True, a model can be given a graphic representation – but so can any idea as long as symbolic or conventional diagrams are allowed. Diagrams, whether representational or symbolic, are meaningless unless attached to some body of theory. On the other hand theories are in no need of diagrams save for psychological purposes. Let us then keep theoretical models apart from visual analogues. (More on models in Bunge, 1972a.)

9. Upshot

In factual science analogy and analogical inference are welcome as theory construction tools. By the same token they are signs of growth, symptoms that the theory is still in the making rather than mature. A mature classical electrodynamics has no need for elastic tubes of force: the field – a nonmechanical substance – suffices for all purposes, the mechanical analogues being regarded as removable appendages. Likewise a mature quantum electrodynamics will need no virtual photons jumping out of electrons and back: it will take the Feynman diagrams as mnemonic devices attached to a computation method rather than as literal descriptions (Bunge, 1955).

When a factual theory attains maturity it involves only literal interpretations and it yields only literal explanations: it does not involve *as ifs*. Surely scientific explanation, if deep, is more than just deduction from laws and data: it will be a subsumption under generalities all right but, among these, some mechanism hypotheses will figure – that is, some assumptions will go beyond external or input-output relations. Yet such explanations in depth, or interpretive explanations (Bunge, 1967c, 1968a), are alien to metaphorical explanations, which are superficial for being limited to similarities and for failing to concern the real thing. Correspondingly the metaphor view of scientific explanation, which has recently been commended in place of the deductive account (Hesse, 1966), is totally inadequate.

It behoves the foundational worker and the philosopher of science (i) to recognise the splendid gifts analogy statements and analogical inferences can make to theory construction, (ii) to analyse them rather than describing them in metaphorical terms, (iii) to expose the uncritical use of analogies and arguments from analogy in science, (iv) to distinguish the constitutive assumptions from the purely heuristic ones (as Kant did two centuries ago), and (v) to help science get rid of its heuristic scaffolding for, if kept beyond the initial construction stage, it will eventually hamper the further growth and clarification of the theoretical edifice, as is right now happening with some of the classical analogues that linger in the quantum theories long after having exhausted their heuristic power. Now, the best way of clearing the heuristic scaffolding and getting hold of the real assumptions (both overt and covert) of a theory is to exhibit its axiomatic foundations. Let us then turn to physical axiomatics.

NOTE

* Some paragraphs are reproduced from Bunge (1967f) with the permission of the editor and publisher.

CHAPTER 7

THE AXIOMATIC FORMAT

1. THE THREE AVENUES OF APPROACH
TO PHYSICAL THEORY

A scientific theory can be expounded in either of three ways: historically, heuristically, or axiomatically. A historical exposition, if faithful and insightful, will display the original problem situation, the various attempts made to cope with it – including the wrong moves –, the way the adopted solution was justified, and its actual or likely influence on future developments. Needless to say, although the history of physics is growing fast, very few historical studies perform all of the preceding tasks. The heuristic approach, on the other hand, seizes on the most useful (not necessarily the most fundamental) formulas of the theory and it rushes on to work out their consequences and apply them. It is the approach adopted in the vast majority of publications and courses in physics. While both the historical and the heuristic formats are enlightening, the former is impractical: if a short statement or a quick command of the subject are desired, then the heuristic approach is preferable.

However, the historical and the heuristic approaches fail to exhibit the theory as a whole: they keep silent about most of the presuppositions of the theory, they do not display all of its basic premises, they leave the logical structure of the theory largely in the dark, and they are ambiguous, if not outright inconsistent, concerning its physical meaning. Every conscientious teacher and every careful student feel dissatisfied with those two unsystematic approaches and try to supply some of the missing premises, or to bring some rigor to this or that proof, or to sharpen the physical interpretation of some of the symbols concerned. Hence the bewildering variety of formulations and the corresponding wealth of textbooks and review articles. Much as these attempts at clarification and reconstruction may succeed in particular points, they are bound to be at best partially successful, for they are not systematic enough – they do not remodel the theory as a whole – and systemacity is of the essence of any theory. Indeed, from a logical point of view a theory is, by definition, a system and, more precisely, a hypothetico-deductive system, i.e. one that starts from a definite bunch of hypotheses and proceeds deductively. (Recall Chapter 3.) And systems are not tampered with – except to change them deliberately into alternative systems.

If a more accurate formulation and therefore a fuller and deeper understanding of a theory are sought, either for pedagogical purposes or with a view to improving the theory – or simply for one's private intellectual delight – then the axiomatic approach will be preferred. Indeed, this is the one that gives a global account of a theory and the one that focuses on its essential ingredients instead of distracting us with sidelines, applications, or historical and psychological zig-zags. The axiomatic approach makes a beeline to the core of every theory. Moreover, by its very nature an axiom system cannot be loaded with details: the latter are left to the applications.

However, there is no conflict among the three approaches to theory exposition: each illumines a different facet of a complex object and each has its own goal. The first is interested in the biography of a theory, the second in its capabilities and performance, and the third in what may be called its character: its foundations, both as to structure and content. Therefore it would be mistaken to claim that any of the three approaches is absolutely superior to either or both the other two. The three approaches are mutually complementary and consequently a well-rounded scientific education, even though centered on the heuristic or intuitive format, should give an idea of the extremes of historical chaos and axiomatic orderliness.

I shall not defend the heuristic approach, for it is employed universally. Nor shall I advocate the historical approach, as every educated scientific specialist likes to do some reading into the history of his subject. Instead, I shall advocate the unpopular cause of axiomatics, which is widely misunderstood and hardly practised outside mathematics. (See however Woodger, 1937; Carnap, 1958; Henkin *et al.*, 1959; Bunge, 1967a, 1967b, and Suppes, 1969.)

2. THE LESSONS OF EUCLID, HILBERT, AND GÖDEL

Before Euclid (fl.300 B. C.) there had been views and doctrines but, so far as the extant documentary evidence goes, there had been hardly any theories in the modern sense, with the lonely exception of Eudoxus' theory of proportions. That is, there had been more or less loosely connected statements rather than hypothetico-deductive systems, i.e. systems based on explicitly stated initial assumptions (also called axioms or postulates). True, the notion of a proof had been invented a couple of centuries earlier (probably in the Pythagorean school) and the need for making some assumptions in order to prove anything had been recognised alongside the concept of logical consequence. But the proofs were as isolated as the premises themselves: neither of them were systemic. It had been the happy reign of the problem-solver rather than that of the theory-builder: problems were attacked one after another and they were solved by grabbing any premises that looked promising, without caring much for either circularity or consistency, let alone for homogeneity (the belonging to a set of mutually related items).

The system of geometry which Euclid expounded in his *Elements* was not just a heap of computation recipes (like most of Sumerian and Egyptian mathematics). Euclid's system was even more than a huge collection of pieces of mathematical knowledge: it was probably the first full-fledged theory ever invented by mankind. To belittle Euclid by saying that he "only" codified the mathematical knowledge of his day is to betray a poor understanding of the nature and importance of theories. Moreover, Euclid formulated his geometrical theory in the most complete and cogent way possible at that time – namely in the axiomatic fashion he himself introduced. Euclid's lesson in the methodology of theory construction was then this: *If you care for systemacity and rigor, try the axiomatic method.*

Though always admired as a paradigm of theory formulation – to the extent that Spinoza, Newton, and many others tried to formulate their main theories *more geometrico*, i.e. axiomatically – axiomatics lay largely fallow until the turn of our century. It was revived at that time by two groups of mathematicians: those who began to realise the formal similarities among a number of different theories (thus originating the concepts of abstract theory and of models thereof), and those who worried about the traps that lay hidden in the intuitive or heuristic formulations of certain theories – first the calculus, then set theory. Axiomatics was brought back to life, in short, as a tool for unification, clarification, and cleansing.

The two most influential axiom systems of that initial period of modern axiomatics were, of course, G. Peano's axiomatic reformulation of Dedekind's postulates for the natural number system (1889), and D. Hilbert's *Grundlagen der Geometrie* (1899). The latter both vindicated Euclid's method and overhauled it, by keeping a more rigorous control on the assumptions and derivations – among other things by dispensing with drawings, which exclusion forced him to state all premises in an explicit symbolic form. Quite apart from its intrinsic mathematical significance, Hilbert's metatheoretical lesson was this: *No axiom system is final: it is always possible, in principle, to dig out deeper (stronger) axiom layers* (Hilbert, 1918).

Since that time most theory builders have regarded axiomatics as the ideal format for mathematical theories, particularly in algebra, but increasingly so in other branches of mathematics as well. The ideal format but not the perfect one: it is not only that, as Hilbert saw, every period supplies the materials for building the best axiom system in a given field, but that no rich theory can be perfect even if expounded axiomatically. Indeed, Gödel (1931) proved that no consistent axiom system involving arithmetic can contain every formula in the field it intends to systematise: every such system is necessarily incomplete if it is consistent. (And, if consistent and complete, the system is not fully axiomatisable.) True, a stronger axiom system can always be built, as Hilbert had enjoined us to do, that will encompass more statements than the previous theory. But even so it will remain incomplete: perfection, then, is not to be equated to attainable ideal. Gödel's lesson concerning theory construction was, in a nutshell, this: *There can be no perfect axiom system: all we can and must strive for are ever improved axiom systems.*

Every particular axiom system is then limited, but there is no a priori limitation to the sequence of progressively better axiom systems. Axiomatics is imperfect but no better kind of theory formulation is in sight, so that to renounce axiomatics for its limitations is like proposing to discontinue the human lineage because no individual can be perfect. Besides, imperfection is necessary for progress.

3. THE CURRENT STATE OF THE ART IN PHYSICS

Hilbert was not only one of the greatest mathematicians and logicians in history but also a distinguished theoretical physicist and a champion of the use of axiomatics in the whole of science. In 1900 he challenged the International Congress of Mathematics in Paris by posing twenty-three hard open problems. Most of them have since been solved, some only very recently. But Hilbert's celebrated 6th problem, namely to axiomatise physical theories, remains largely open. Hilbert himself devoted some effort to this task by axiomatising the elementary or phenomenological theory of radiation (Hilbert, 1912, 1913, 1914) and his own unified field theory of gravitation and electromagnetism (Hilbert, 1924). Unfortunately he chose the wrong subjects: the former theory had been under revision since Planck's quantum *putsch* (the quantum revolution came much later), and the second theory was premature. So, Hilbert's essays in physical axiomatics were paid hardly any attention to. With a few exceptions, physicists have continued to formulate theories in a casual way, sometimes even in a confused way.

There have been some exceptions, though. The best known, or at least the most quoted one, is Carathéodory's rather incomplete formulation of thermostatics (Carathéodory, 1909). It has hardly been used due to its unintuitive character. Besides, it made the usual mistake of seeking physical meaning in experiments. This is mistaken if only because (*a*) usually it is some of the theorems, not the postulates, that come close enough to experiments, (*b*) the design and interpretation of an experiment involves not only the very theory concerned but also a number of other (auxiliary) theories capable of handling the various traits of the experimental arrangement (see Chapter 10); (*c*) all these theories must be at hand, equipped with a more or less precise content, before they can be applied; (*d*) experiments are not expected to disclose meanings anyhow – but rather to yield data capable of checking theories, applying them, and firing new questions, and (*e*) the problem of meaning starts with the building blocks (undefined concepts) of the theory: these are the ones that must be supplied with a content to begin with if the compound constructs are to get one. In short, Carathéodory's first essay in physical axiomatics was marred by the untenable philosophy of operationalism.

The next best known essays in physical axiomatics were Carathéodory's (Carathéodory, 1924) and Reichenbach's (Reichenbach, 1924) alleged axiomatisations of special relativity. They had similar assumptions and they were equally unsuccessful. In particular, both of them overlooked Maxwell's electromagnetic field theory, without which special relativity makes little if any sense, if only because it happens to be about events connectible by electromagnetic disturbances. Those two formulations took the objects of the theory to be material points rather than unspecified physical systems and reference frames immersed in an electromagnetic field; both assumed that every material point could lodge an observer capable of sending off and receiving light signals at will and without recoil; and none of these axiom systems entailed the Lorentz-transformation formulas. Yet Reichenbach's pseudoaxiomatics still still passes for perfect or nearly so among philosophers.

Another famous attempt was von Neumann's in relation to quantum mechanics. In his epoch-making book (von Neumann, 1932), which enriched the mathematical framework of the theory, von Neumann is wrongly supposed to have laid down the axiomatic foundations of quantum mechanics. As a matter of fact his exposition lacks all the characteristics of modern axiomatics: it does not disclose the presuppositions, it does not identify the basic concepts of the theory, it does not list all the initial assumptions (axioms), it fails to propose a consistent physical interpretation of the formalism, and it is ridden with inconsistencies and philosophical naïvetés. (See Chapters 4 and 5.) Yet for some strange reason it passes for a model of physical axiomatics. But then this is not the first case: also Mach's formulation of classical particle mechanics (Mach, 1942) often passes for an axiomatisation of it even though Mach never came near axiomatics and was moreover suspicious of every theory, for what counted for him was fact alone (see Bunge, 1966).

In the meantime the study of axiom systems made dramatic strides. Whole new disciplines – metamathematics and model theory – were born and made important contribution to the theory of theories between the two world wars, and quickened considerably their pace after the last. The first essay in physical axiomatics which took stock of some of these advances was probably the work of McKinsey et al. on classical particle mechanics (McKinsey, 1953). This was the first time the primitive concepts of a physical theory were pinned down and characterised, the first time most of the necessary axioms of the theory were formulated, and the first time they were subjected to the metamathematical tests of consistency and independence (the latter both at the level of concepts and at the level of statements). Next came Noll's axiomatisation of a much richer and realistic theory, namely classical continuum mechanics (Noll, 1959). His work placed special emphasis on mathematical modernity and rigor, and availed itself of functional analysis. A third important step was taken by Edelen (Edelen, 1962), who axiomatised an entire class of classical field theories. These three works have been pioneers in their fields and they have shown, among other things, how axiomatics can help shape up and extend classical physics, in addition to bringing a deeper understanding of it. Yet curiously enough none of them has found its way into the physics textbooks: we prefer to teach old-fashioned mathematics and to ignore logic – and sometimes even entire chapters of physics, such as continuum mechanics.

The works mentioned so far were due to mathematicians or logicians. Consequently, with the exception of Hilbert's, they did not care much for physical content. Or, if they did pay some attention to it, they often borrowed uncritically the physicist's operationalist doctrine of meaning, according to which testability is necessary for meaningfulness rather than the other way around.

The recent contributions of physicists to physical axiomatics include, first and foremost, the work of Wightman and his school on quantum field theory. This research, partly motivated by the well-known in-consistencies (e.g. divergences) of the usual formulations of quantum electrodynamics (Wightman, 1956), has been curiously misunderstood. Some dislike its attempt at mathematical rigor; others reproach it for not having predicted any new "effect"; others object to it because it pos-tulates the existence of an unobservable entity – a field – and finally others because they seem to think that 'axiomatic' means a priori, independent of experience, and consequently adamant to it. Few take axiomatic field theory for what it is: an attempt "to analyse the general notions which underlie all relativistic quantum field theories" (Jost, 1965) without any claims to finality or incorrigibility.

This brief and incomplete survey of physical axiomatics should suf-fice to show its present immaturity. This state of affairs, in sad contrast to the flourishing of axiomatics in mathematics, recommends promoting the axiomatic approach in physics if only to see what comes out of it. The rest of this chapter will be concerned with suggesting how this project – the axiomatic overhauling of theoretical physics – can best be implemented.

4. GENERAL CHARACTERISTICS OF AXIOMATICS

The difference between an intuitive or heuristic presentation of a theory, and an axiomatic formulation of it, is similar to the difference between the actual course of a piece of research and a final report on it. It is a difference in systemacity and in priorities and therefore in clarity as well. This is all there is to axiomatics, however difficult its implementation may prove to be in most cases. Indeed, *to axiomatise a body of knowledge is just to exhibit its main ideas in an orderly fashion.*

Now, the term 'idea' covers both concepts and statements. A concept, such as the one of electric field strength, is a unit entering some statement, such as "The field strength dies off with increasing distance." Consequently, the axiomatisation of a theory consists in drawing an orderly list of both the main concepts and the main statements of the theory, in such a way that all the remaining concepts and statements of it be derivable from those main ideas.

A main or basic idea is one serving to build further ideas by purely logical or mathematical means: it is either a concept used (usually with the cooperation of other concepts) to define some other concepts, or a statement used (usually jointly with other statements) to deduce some further statements. The basic concepts of a theory are called its *primitive* or *undefined* concepts, while the basic statements in it are called the *axioms* or *postulates* of the theory. (No difference is made between 'axiom' and 'postulate' in modern axiomatics.) Thus the four concepts of field intensity and induction (or, equivalently, the two corresponding field tensors) are basic or undefined in Maxwell's own version of classical electromagnetism, while Maxwell's equations are among the axioms of this theory. All this means is that, in this particular formulation of Maxwell's theory, those happen to be among the basic units out of which the whole theory is built. The concept of energy density, no doubt important, is not logically basic because it can be defined in terms of those primitives; and the concept of total energy is even more derivative, as it can be defined in terms of the energy density. Likewise, the law of conservation of the field energy is important but not basic, as it constitutes a theorem entailed by those axioms.

However high, the status of primitive and the one of axiom is not absolute but context-dependent. Thus, the concept of number is a primitive in elementary arithmetic just as the concept of force is one of the primitives of Newtonian particle mechanics. But in alternative theories numbers are defined in terms of sets, and the force concept is often definable in terms of the concept of potential. Accordingly certain statements are degraded, in alternative theories, from the rank of axioms to that of theorems. This contextuality or relativity of the status of basic idea is welcome, for is allows us to search for ever stronger and deeper ideas: for richer concepts and richer postulates, out of which the previous basic ideas can be derived. Consequently the sentence 'The concept (or the axiom) A is basic' should be relativised to 'The concept (or the axiom) A is basic in the theory T'.

It might be thought, though, that since all the concepts and all the statements of a theory are or may be needed, all of them are equally important: that the working scientist must prefer the democracy of ideas to their stratification into first-class and second-class ideas (concepts and statements). This is indeed so from a semantic point of view, i.e. as regards meaning and truth. Thus, the concept of field flux is no less important than the concept of field strength even though the former is definable in terms of the latter; moreover, fluxes are more important than intensities in experimental physics. And the field equations are no less important than the variational principles that entail the former and which, after all, may be regarded as mathematical and heuristic conveniences. But all this is beside the point: for all its physical importance, a derived idea (a defined concept or a derived statement) is logically secondary. The basic/derived distinction is a purely logical one. Moreover, it is a useful distinction even though it is relative to the context, for it enjoins us to look for the central ideas, to spot the logically strongest ideas, and it helps us avoid running in circles. Thus anyone conversant with this distinction will abstain from attempting to define every concept (definitionism) and to prove every statement (demonstrationism). More in this in Chapter 8.

We are now in a position to give a somewhat more precise characterisation of axiomatics. To axiomatise a theory is to lay down a set of initial assumptions such that these (*a*) give a sufficient characterisation of all the basic concepts of the theory, and (*b*) yield all the standard statements (formulas) of the theory. But before writing down any such assumption we need a language to express it. Therefore we must start by casting a glance at the logical and mathematical background of a physical theory.

5. THE FORMAL BACKGROUND

There is a single theory that starts from scratch: mathematical logic (which is actually a set of theories). Indeed, the truths of logic or tautologies – such as "$A\Rightarrow(B\Rightarrow A)$" – are those that can be proved without resorting to assumptions other than the rules of logic. All other theories presuppose at least logic and usually a lot more. More precisely, the least a mathematical or a scientific theory takes for granted is the so-called ordinary (two-valued) predicate calculus enriched with the microtheory of identity. This theory is necessary and sufficient to analyse the concepts, formulas, and reasonings occuring in mathematics and in science – or rather to analyse their form. In fact, every statement in mathematics or in science is, as far as its form is concerned, a formula of that calculus; and every valid reasoning is an instance of an inference pattern consecrated by that same theory. It is not that actual scientific reasoning needs logic in order to keep going: logic is not supposed to construct anything outside itself, but to control the validity of whatever is being built with concepts. Nor is it implied that mathematical and scientific research avoid analogical and inductive arguments, which can be fruitful even though they are always invalid. What is implied is that no theory can be expected to validate nondeductive reasonings, even though their conclusions may be true.

In addition to the ordinary predicate calculus there are many other logical theories, such as model logic and many-valued logics, but the logic that is built into most of mathematics and all of physics is the former. (The exception is intuitionistic mathematics, which covers only a small portion of mathematics and is therefore of no interest to physics.) This platitude had to be recalled in view of the claim that quantum theory uses a logic of its own, in which the distributive laws for propositions fail. If this were true, then the quantum theory would have an entirely different mathematical formalism and it would be hardly possible to join it to classical physics, as we in fact do whenever we fail to quantise an external field or when we enter some experimental result – e.g. the value of a constant – gotten with the help of measurements designed and interpreted with the help of classical physics. In short, not even the quantum revolution has changed our logic (Bunge, 1967a; Popper, 1968; Fine, 1968).

Ordinary logic is, then, the very first component of the formal background of any physical theory. Its second and last component is of course the collection of mathematical theories actually employed in the given physical theory. In turn, the mathematical basis or background of all the mathematical theories so far employed in physics is set theory. (The exceptions are the so-called elementary theories, such as the elementary theory of groups.) Indeed, almost every mathematical concept can be defined, in however devious a way, in terms of the basic notions of set and set membership. But, for better or for worse, just as in the case of logic there are several inequivalent set theories. Which shall be chosen for building physical theories? Surely most physicists do not face this *embarras de choix*, but the foundational worker does. Indeed, it is not just that he would like to adopt the version of set theory that is mathematically the most satisfactory (supposing there is one). The worker in the foundations of physics is also concerned with choosing the set theory enabling him to elucidate physical concepts in the most natural way. Thus if he is dealing with some characteristic of a multi-component system, such a light beam traversing a transparent medium, he may have to conceive of the system as an ordered couple (if, indeed, order is essential), so that the characteristic in question will be formalised as a function on the cartesian product of 2 sets. (For instance, the refractive index is a joint property of light and optical medim, and it may be analysed as a real valued function on the set of couples light beam-point in an optical medium.) But then, since the compound system is as much of a thing as every one of its components, the foundational worker may feel dissatisfied with the standard construal of an ordered n-tuple as a set of sets: he may prefer to interpret it as an individual on its own right. Therefore, among all competing axiomatic set theories, he will probably prefer Bourbaki's version.

In short, physical axiomatics presupposes both logic and mathematics, or rather those portions of logic and mathematics that have heretofore been exploited in physics. This formal apparatus is necessary for the orderly reconstruction of physical theories but it does not suffice to this end, for it remains silent concerning physical meaning. We must therefore supply further items to the background of a physical theory.

6. THE PHILOSOPHICAL BACKGROUND

Scientific research involves certain ideas that are elucidated neither in formal science (logic and mathematics) nor by any special empirical science. In particular, the axiomatic reconstruction of a scientific theory presupposes a number of concepts and hypotheses which, though far from being purely formal or syntactical, are much too general and important to be the private property of any particular science. These are certain philosophical ideas, such as those of meaning and truth, and certain protophysical ideas, such as those of system and time. Let us start with the former, leaving protophysics for the next Section.

Consider the following statements:

S1 The value $X(\pi, k, t)$ of the material coordinate function X *represents* the position of the particle π in (or relative to) the frame k at the instant t.

S2 For every particle π there is a septuple $\mu = \langle M, X, P \rangle$ *called* a mass point, where M *designates* the mass, X the position, and P the momentum of π, such that $\mu \mathrel{\triangle} \pi$ (read 'μ *models* π').

S3 The point mass model *holds approximately* for classical particles but it is *nearly false* for quantum-mechanical systems.

None of the italicised words belongs to logic, mathematics, or factual science: they are all *semantical terms*. And all three statements are *semantical statements*, for they concern the meaning of certain symbols or the truth of certain ideas. Semantics, in turn, is an old branch of philosophy – but one which has made more dramatic strides ever since, at the beginning of this century, it was taken up by logicians and mathematicians such as B. Russell, R. Carnap, and A. Tarski and their followers.

The above statement *S1* performs two jobs. On the one hand it functions as a rule of designation or naming, by pointing to the physical counterpart of a symbol. On the other hand *S1* goes beyond a mere linguistic convention, for it expresses the idea that the position of a particle is fully described or represented by a vector valued function on the set $\Pi \times K \times T$ of triples: particle-frame-instant. This assumption might be false: it was not made when people believed in absolute space. Moreover, the assumption makes no sense to operationalism, which would require K to stand for the set of observers rather than for the set of physical frames of reference, and would replace the concept of position by the concept of measured value of the position. Finally, $S1$ is pointless in quantum mechanics except for the average position of a system. In short, although $S1$ is a semantical assumption, it is not totally conventional – it is a hypothesis rather than a rule. And in any case it is neither a mathematical nor a physical assumption but a semantical one. We shall argue in Sections 9 and 10 that physical axiom systems cannot fail to contain such semantical assumptions at their very basis.

The second semantical assumption, $S2$, is the conjunction of a definition (the one of the mass point concept), three rules of designation (one for every coordinate of the 7-tuple concerned), and one assumption – namely the hypothesis that a mass point represents a particle rather than, say, a field, or a measurement act, or an item of information.

Finally, the third statement, $S3$, summarises the performance of the mass point model in terms of its degree of truth. Unlike $S1$ and $S2$, this one is not an assumption but a result of a critical examination of particle mechanics as regards its ability to represent truthfully the vicissitudes of physical particles. However, the study of statements of the kind $S3$ concerns not only physics (or rather the methodology and metatheory of physics) but also semantics, for these statements involve the semantical notion of partial factual truth. (For the semantics of science see Bunge, 1972b.)

Consider next these other statements:

$M1$ There are *physical objects* – i.e., objects whose existence and properties do not depend on being perceived, thought of, or measured.

$M2$ Every physical object fits into some set of *physical laws* – i.e., stable objective patterns.

$M3$ It is possible to *know*, though only tentatively, approximately, and gradually, the physical laws as well as some of the idiosyncrasies of individual physical objects.

These three statements are of a *metaphysical* nature: they make claims about the existence of the external world, its lawfulness, and its knowability. None of these statements is specific enough to be dealt with by a given physical theory. None of them can be disproved either theoretically or empirically: they can only be confirmed – indeed, every successful research project reinforces the above metaphysical hypotheses while no failure would be blamed on them. Nay, the latter are necessary conditions for the very launching of any research enterprise in factual science. For, if there were no physical objects, we would not bother to discover them; if they behaved in an erratic fashion we would not try to find out their laws; and if they were impregnable to our cognitive effort we would not waste our time on it. In short, the above are *metaphysical presuppositions* of physical research. Furthermore, specific instances of $M1$ – e.g., "There are gravitational fields" – belong to the axiomatic reconstruction of physical theories, as we shall see in Chapter 8.

In sum, both research and the finished products of research in the physical sciences presuppose a number of metaphysical hypotheses, all of which deserve being discovered, collected, systematised and scrutinised in what may be called the metaphysics of science. The fact that science presupposes some metaphysics need not disturb us but should rather motivate the research for cogent, clear and fertile systems of metaphysics – and moreover systems capable of helping scientific research rather that hindering it with overcautious prohibitions or overpermissive principles. If we do not perform this task then the metaphysicians of a traditional or speculative cast of mind will bungle it. In any case, there is no escape from metaphysics. (More on this in Bunge, 1972a.)

In conclusion, physical theories presuppose and involve a number of semantical and metaphysical ideas. True, most of these ideas have not been sufficiently elucidated: the semantics and the metaphysics of physics are still half-baked. But this is one more reason for keeping the oven hot – as long as the oven is equipped with light and window.

7. Protophysics

In addition to the formal and the philosophical components of the background of a physical theory we can discern a third component of it: protophysics. This field is made up of a number of principles and theories concerning very general traits of physical systems – so pervasive, in fact, that protophysics qualifies as a special branch of exact metaphysics.

One of the theories of protophysics is mereology, or the elementary part of the general theory of systems. The aim of mereology is to elucidate in a systematic fashion, i.e. by setting up a theory, or rather a set of theories, the basic notions of system and of being a part of a system – or of constituting a system either by juxtaposition (physical addition $+$) or by interpenetration or superposition (physical product $\dot{\times}$). It is obvious that some such theory is needed to introduce in a clear way the nonemergent or hereditary properties of wholes. Thus the electric charge $Q(x+y)$ of a system composed of two bodies x and y will be the sum of their individual charges, while the total energy of two superposed linear fields – i.e. of the compound system $x \dot{\times} y$ – equals the sum of their individual energies. A possible elucidation of the basic concepts of mereology is provided by a physical interpretation of a Boolean algebra (Bunge, 1967a); another is a model of ring theory (Bunge, 1972a).

Once clear concepts of a simple and of a complex system are available, we can introduce the concept of physical property, elucidating it as a map (function or operator) on the set of all systems of a kind. In this way the ghost-like notion of a system-less or unsupported property is avoided and, together with it, the misconstrual of quantitative properties as numbers goes. We may thence proceed to define the notion of state of a system as a point in a certain space (the state space) associated with it. It then becomes possible to represent an event as an ordered couple of points in the state space. Clearly, a process will then be definable as a sequence of events, i.e., as a trajectory in the state space (Bunge, 1968b).

After the concepts of thing (system) and its changes (events) have been elucidated we may proceed to construct several relational ideas of physical space and time. The best way to do it is, of course, to proceed axiomatically. Thus a metrical physical geometry can be built by specifying conditions for a distance function defined on the cartesian product of the set of physical points by itself – a physical point being characterised as one at which not every physical property vanishes. Likewise the foundations of a theory of time may consist of a bunch of axioms for the duration, construed as a real valued function on the set of triples: event-event-reference frame (Bunge, 1968d).

As with system, property, event, space, and time, so with other protophysical (or metaphysical) concepts, such as those of causation and physical probability. They all can be elucidated with the help of a few logical and mathematical notions, and their elucidation provides a clear (but by no means perfect) protophysical foundation of physics. True, physics did not wait for such an elucidation to start; moreover, proto-physics could not have begun until physics was well under way. Like-wise the calculus was started long before it could be given a rigorous – or rather a far more rigorous yet not definitive – foundation on set theory, algebra and topology. However, such a foundational work was eventually needed in order to avert catastrophes, bring in unity, and thereby facilitate the development of analysis. Even if protophysics were to bring nothing but some clarification, it should be welcome or at least tolerated.

It is then clear that every physical theory presupposes a number of ideas that are not the concern of physics: (*a*) the formal background (logic and mathematics), (*b*) the philosophical background (semantics and metaphysics), and (*c*) protophysics (basic systems theory, general theories of space and time, physical probability theory, etc.).

Some physical theories presuppose also one or more specific physical theories. Thus solid state physics takes quantum mechanics, classical electromagnetic theory, and statistical mechanics for granted: it is an application of these theories, which may in turn be regarded as funda-mental in several degrees. A strictly *fundamental* physical theory is not one that is free from presuppositions but one that presupposes no other physical theory. Classical mechanics and quantum electrodynamics are examples of fundamental theories. The point of making such a distinc-tion between fundamental and nonfundamental theories is to avoid running around circles, like trying to deduce electromagnetic theory from Coulomb's law in conjunction with special relativity, which in fact is based on Maxwell's theory.

This closes our quick review of the background of a physical theory. Let us now proceed to examine the notion of a primitive concept.

8. PRIMITIVES

The building blocks of an axiomatic theory are, of course, its undefined or primitive concepts. (Recall Section 4.) The primitives employed in building an axiom system may be classed into *generic* and *specific*. The former enter a number of theories belonging to different fields while the latter characterise the particular theory concerned. Thus the generic notion of state and the specific concept of entropy are undefined concepts of thermodynamics. A physical theory can borrow its generic primitives from its background in the hope that someone will cultivate this land. In any case, a particular physical theory is not expected to elucidate any generic physical concept.

The starting point of a physical axiom system is then a set of *specific primitives*, i.e. of undefined concepts that concern the particular kind of system studied by the theory and are consequently not elucidated by any of the theories that make up the formal, the philosophical, or the protophysical background of the particular theory in question. However, it is often convenient to shift some of these protophysical concepts to the foreground and treat them on a par with the specific primitives of the theory. This is the case with the concepts of space and time. There are three good reasons for including these concepts in the primitive base of any physical theory that happens to employ them. First, for the sake of uniformity: in this way all the "variables" (actually sets and functions) of the theory which are assigned a physical meaning are listed and elucidated to some extent, and so regimented and watched over. Second, because different theories may need different concepts of space and time – and a few none at all. (Statics is a timeless theory while the elementary theory of electrical networks is a spaceless theory.) Third, because protophysics is largely a no-man's land, so that protophysical concepts are often shrouded by a fog that can be lifted only if we scrutinise them often enough.

Let us then agree to call the *primitive base* of a physical theory the bunch of undefined concepts of it that are assigned a physical meaning and which occur in the physical assumptions of the theory. Thus the primitive base of geometrical optics consists of three sets and one function: Euclidean three-space, light rays, transmitting media, and refractive index. The function of the axioms of geometrical optics is to characterise, both formally and semantically, all four primitives and to glue them together into the basic law of the theory, namely Fermat's principle. And such a characterisation (not definition) of both the fundamental concepts and the fundamental statements of a physical theory has the same ultimate aim as the corresponding preaxiomatic (or naive) formulation, namely to describe certain physical entities and explain and predict their behaviour. Whether axiomatic or nonaxiomatic, then, the purpose of a physical theory is different from the purpose of a theory in pure mathematics: where the former describes the latter defines. Thus while the axioms of lattice theory define and even create the whole category of lattices, the axioms of geometrical optics attempt to mirror things that cannot be defined, namely optical systems. At this point, then, physical axiomatics parts company with mathematical axiomatics.

Formalists would claim that, since mathematics is based on set theory, only set theoretical concepts need be considered in building a scientific theory. And furthermore, since every physical concept has the structure of a set theoretical object, they would hold that there are no physical primitives and, consequently, no differences between a mathematical theory and a physical one (Suppes, 1967). On this view a physical theory with only two basic concepts, such as a set X and a function F on and into X, would have no primitives of its own, since set theory takes care of both X and F. But this is like holding that mathematics has no business of its own for, after all, every mathematical formula is a formula of the predicate calculus. Set theory characterises certain very general notions, such as the one of function, whereas specific mathematical theories characterise specific notions, such as the ones of additive measure and of sine function, about which set theory knows nothing in particular. Something similar holds, a fortiori, for the basic concepts of physics: even though mathematics is expected to disclose their formal structure, their physical meaning overflows mathematics and must be specified by physics. Surely the (nonrelativistic) concepts of mass and electric charge are mathematically identical, and so are those of spatial and temporal coordinates, but their content is definitely different. In conclusion (a) formalism does not render justice to physical concepts, which are not empty forms, whence (b) unlike a mathematical theory, a physical one requires semantical assumptions relating its symbols to some of the entities in physical reality and their properties. (Recall Chapters 3 and 4. More on formalism in Bunge, 1972a.)

9. Axioms

The problem of characterising a basic or undefined physical concept is then two-fold: both the form and the content of the concept should be specified or at least outlined. In an axiomatic theory such a specification is a task for the axioms. The axioms should determine the mathematical status of every primitive (set, differentiable manifold, Hilbert space, or what not), they should sketch the physical content of it, and they should relate every primitive to some other concepts of the theory in such a way that the main traits of the physical system concerned be accounted for.

The axioms of a physical theory should then fulfil three functions: a formal or mathematical one, a semantical function, and a properly physical function. In other words, every well built physical axiom system will contain postulates of three (and only three) kinds:

(1)   *Formal* (mathematical) assumptions, or *FA*'s for short – e.g., "*P* is a probability measure on the set S^2 of all the ordered pairs of elements of the set *S*".

(2)   *Semantical* (meaning) assumptions, or *SA*'s – e.g., "If the ordered pair $\langle s, s' \rangle$ is in S^2, where *S* is the state space of the system, then $P(\langle s, s' \rangle)$ stands for (represents) the strength of the tendency of the system to go from state *s* to state *s'*."

(3)   *Physical* assumptions, or *PA*'s – e.g., "$P(\langle s, s' \rangle) = P(\langle s', s \rangle)$".

Of the three groups of axioms the third one, constituted by the physical assumptions, forms the core of any physical theory. Indeed, while the formal axioms concern the form of the basic concepts and the semantical axioms take care of their meaning, the physical assumptions are about the physical systems themselves – which, after all, are the *raison d'être* of the theory. In turn, by far the most important amongst the physical axioms are those intending to represent objective physical laws. The remaining physical assumptions, in particular constraint formulas and boundary value conditions, even though logically independent of the law statements are subsidiary to the latter: they are just further restrictions on the various variables and functions interlinked by the law statements. An axiom system which does not have at least one law statement fails to qualify as a physical theory.

Which physical assumptions should be postulated in a theory? Clearly, all those and only those formulas which cannot be proved within the given theory and yet are assumed or at least hoped to be true to some approximation. One will not postulate an equation of motion, or a field equation, if one can derive it from some stronger assumption, e.g. a variational principle, particularly if this stronger axiom entails further consequences as well – e.g. conservation equations. Needless to say, by 'derivation' or 'proof' we mean a purely conceptual operation – a reasoning – whereby the desired conclusion follows from a set of premises by virtue of the rules of inference of the underlying logical and mathematical theories.

In particular, we shall not set out to "prove" within, say, a quark theory, that there are quarks: such an existence "proof," or rather intimation, is supplied by showing that there is something out there that seems to satisfy certain assumptions of a quark theory. And this process of verification involves the theory concerned, allied to several other theories, but does not belong to it. Nevertheless, if and when such an empirical test has been passed, certain theorems of the theory may point to the possible existence of hitherto unsuspected properties of the theory's referents. In short, *in* a theory only theorems can be proved or disproved. Whatever can be proved or rendered plausible *about* a physical theory is so from outside the theory: either with the help of experiment, or within the given theory's metatheory – as is the case with a consistency proof.

In any case, if we wish to prove much we must make strong assumptions. (More on this in Chapter 8, Section 3.) This will not commit us to believing them all, or even any of them: it will commit us only to the logical consequences of those assumptions provide we assert the latter. Should any of those consequences turn out to conflict with other accepted ideas (data or theories), we would give up some of the axioms without qualms – precisely those axioms entailing the wrong consequences. This is, of course, a normative rather than a descriptive statement: we often behave inconsistently, by making repairs at the theorem level (e.g. by introducing last-minute cut-offs) instead of correcting the postulates.

In particular we must make strong existence assumptions, if only to see them refuted by observation. Thus if the theory is about certain entities we like to baptise "freakons," we are bound to start off with the strong hypothesis that there *are* freakons, even if there is no experimental evidence for them. Otherwise, i.e. if the set of freakons is taken to be empty from the start, the theory will be vacuously true, for anything can safely be predicated of nonentities. More precisely, the very first axiom of our freakon theory will have to read something like this: "(*a*) $F \neq \emptyset$.

(b) Every f in F represents a freakon." Once we get testable consequences of such strong axioms we can hope to check them by experiment. Should the latter fail to exhibit (or rather suggest) any freakons, or should it fail to confirm that freakons have the properties assigned to them by the theory, the latter would have to be given up.

But a physical existence assumption, though necessary, is insufficient: it cannot even be put to the test unless the thing that is being hypothesised is given a hypothetical description. That is, we must make precise assumptions about the properties, constitution, and behaviour of the referents of our theory. Such hypotheses are *physical assumptions* proper. If general and corroborated satisfactorily enough they will be called *laws* or, better, *law statements*, for the physical laws themselves are supposed to be the objective patterns those formulas attempt to capture. Thus in the case of our freakons we may postulate, say, that *The more there are the Q-er they become*, where Q is a new property original with freakons. However, this proposed physical assumption is too vague, not only because it involves an obscure new property Q but also because it is couched in ordinary language. We must commit ourselves to definite hypotheses, otherwise we won't be saying anything definite and we won't be able to suggest any evidence definitely in favor or against our hypothesis. Suppose that, among the infinitely many mathematical formulas consistent with the given ordinary language statement, we choose

PA_1 For every f in F: $dQ/dN=aN$,   where a is a positive real.

This is formally quite precise but it is semantically indeterminate, hence empirically untestable, if only because there are a number of quantities that increase with the total population. We must link the new property Q to some well known physical property: only thus shall we be able to estimate Q or even to recognise a freakon. In general: isolated hypotheses are untestable.

Suppose our freakon theory hypothesises a relationship with an old acquaintance, the electromagnetic radiation theory, namely thus: *Freakons feed on photons*. A simple mathematical formulation of this conjecture, in terms of the field energy density ρ, is this:

PA_2 $dN/d\rho=b$,   with b a positive real.

We can now prove some theorems and ask the experimentalist to test for them. But this only because we have tacitly assumed (*a*) certain obvious mathematical properties of our basic functions (e.g. differentiability), and (*b*) a definite assignment of physical meaning to every symbol in our axioms. Now a trait of axiomatics is, precisely, that it leaves nothing tacit, nothing outside the axioms. Hence our initial assumptions have got to be supplemented with two further batches of axioms, one of mathematical, the other of semantical assumptions. We shall take up an analysis of these other kinds of axiom in the next Section but not before we reap some general conclusions.

Our first conclusion is that in axiomatics, just like in ordinary life, something must be risked if anything is to be gained: in other words, we should not shy off from strong axioms as long as they are testable and that they promise to explain something we did not understand before. Second, there is nothing sacred about physical axioms: they are just premises to be checked by their consequences as well as by their ability to fit in with some accepted ideas. Even a satisfactory performance in such tests will not guarantee their eternity, just as success in life does not bring immortality.

10. *FA*'s AND *SA*'s: THE FORGOTTEN INGREDIENTS

Surely the meat of a physical theory is constituted by its physical assumptions. Consequently the goal of any axiomatisation venture should be to express those assumptions in a cogent and perspicuous fashion. But the physical assumptions make no sense unless accompanied by both mathematical and semantical assumptions. Indeed, a formula is just a string of signs unless the mathematical nature of the latter is specified and unless something is said about the things and properties the primitive symbols are supposed to represent, however schematically. Thus in the example at the beginning of the preceding section the third formula would be unintelligible without the preceding statements, whereas in the light of these it "says" that the transition probabilities are symmetrical or reversible. Usually the context supplies such an information but it cannot help doing so in an ambiguous way. Since an aim of axiomatics is to remove ambiguities, every physical axiom system should include an explicit statement of all the formal and semantical characteristics of the primitives concerned. The more complex a theory the more explicitly should it be formulated in order to avoid equivocations. But this is possible only if attention is focused on the axiomatic foundations of the theory.

The semantical assumptions are the weakest components of an axiom system, for they do only a part of their job: they only trace a semantical profile of the primitives instead of conveying their full content with accuracy. Physical meanings are so rich and elusive that they can hardly be captured by a single sentence. Fortunately the other components of a theory – the formal and the physical assumptions – contribute to delineating the meaning of the undefined symbols. Thus in our first example the formal assumption makes it clear that P is a property of pairs rather than of individuals, suggesting thereby that P is not on the same level as, say, the temperature. Indeed, the argument of P is not a variable representing a system but one representing a pair of states of a system, so that P is a sort of second-order property – unlike, say, the body mass, which, being a measure on the set of bodies, is a property of individuals. As to the physical assumption in that example, it justifies the interpretation of P as a transition probability or tendency to changing state – in agreement with the objective or propensity interpretation of probability defended by Poincaré, Smoluchowski and Popper. In general, the mathematical and physical assumptions justify the semantical assumptions, which in turn supply the core meaning of the primitives – without however exhausting their full meaning. Drop any of the three components and a lame duck remains. (The preceding remarks presuppose construing the meaning of a sign as the ordered couple constituted by the connotation and the denotation of the idea the sign stands for: see Bunge, 1972a, 1972b.)

It will be noted that in our first example the symbol '$P(\langle s, s' \rangle)$' was interpreted in terms of the suspect notion of tendency or propensity rather than with the help of the clear notion of relative frequency. A first reason for this preference for the physical concept of tendency over the statistical concept of frequency is the following. The number $P(\langle s, s' \rangle)$ concerns an arbitrary member of a collection of state transitions, while the corresponding relative frequency concerns the whole collection: it is a collective (nonhereditary) property. A second reason for not equating probabilities and frequencies is that the latter do not satisfy exactly the axioms of the probability calculus. Surely if a given probabilistic theory is true, then the observed relative frequencies will approach, in the long run, the calculated probabilities. But the two do not *mean* the same thing. Observed relative frequencies provide numerical *estimates* of probabilities but they are not the only estimators and in any case they do not supply the content of a probability formula, just as clock readings do not supply the meaning of the symbol 't'. The meaning of a physical symbol should not be mistaken for the numerical value of the corresponding magnitude as estimated by observation – if only because the design and interpretation of scientific observations require a number of theories that are already equipped with a meaning.

When meanings are confused with experimental procedures or with their results, the operationist philosophy of science is adopted. If, on the other hand, the meaning of a symbol is construed as the connotation (or set of properties) of the construct it designates, together with its intended or hypothetical reference class, then a realist philosophy can be adopted consistently. While the former philosophy focuses on the scientist and his operations, the latter philosophy focuses on the physical object itself and is therefore congenial with the goal of physical science, which is to find out what the world is like, not what people seem to be doing – a task for the behavioural sciences. In any case the very phrasing of a semantical assumption cannot help betraying a commitment to some theory of meaning or other. Once again we see that, if physical axiomatics is taken seriously, then it cannot elude philosophy – which in turn must take science seriously if it is to be of any help.

We are now ready to look into elementary examples of physical axiomatics, as well as to assess the virtues and limitations of the axiomatic approach. These tasks will be undertaken in the next chapter.

CHAPTER 8

EXAMPLES AND ADVANTAGES OF AXIOMATICS

We shall presently exhibit two comparatively simple specimens of physical axiomatics. It must be emphasised that they are bound to differ in important respects from axiom systems in pure mathematics. Indeed, while the latter *define* whole families of *formal* objects or structures, such as lattices or topological spaces, our axiom systems aim at *characterising* (not defining) species of *concrete* objects, namely physical systems, which supposedly lead an independent existence. Hence while the mathematical axiomatiser spins his web without any concern for the actual world, the physical axiomatiser is earth-bound.

That is, whereas physical axiomatics will borrow whatever mathematical ideas are needed, it cannot ape in every point the axiomatisation style adequate for pure mathematics, which boils down to defining some complex predicate usually built with set theoretical components. Thus it would be wrong to introduce the concept of an electrical network through a stipulation like this:

DEFINITION: The structure $\mathscr{G} = \langle G, T, V, e, i, R, C, L, M \rangle$, where G and T are sets, V, e, and i functions on $G \times T$, R, C, and L functions on G, and M a function on $G \times G$, is an *electrical network* if and only if [here comes a list of axioms characterising the mathematical status and mutual relations of the listed primitives G, T, etc.].

Such an axiomatic definition would qualify as a *mathematical* theory – not a very interesting one though. But it would not qualify as a *physical* theory because it can be satisfied by any number of objects, formal or concrete, whereas electrical circuits are a unique species and moreover one whose members are out there in the world, not in here in the mind. We do not invent physical systems out of the blue, the way mathematicians invent spaces. Unlike a Hilbert space, an electrical network is not built out of set theoretical concepts: it is not defined within, or constructed with the help of, set theory but with an electric power source, wires, etc. The best we can do is to give a truthful *description* of a network with the help of carefully manufactured concepts and statements. Mathematicians, used to play God, may dislike this procedure and demand that we *define* physical systems in purely mathematical terms with no admixture of semantical assumptions linking those terms to external objects. (See e.g. Freudenthal, 1970.) That attitude betrays a Platonist philosophy and an utter misunderstanding of the very aim of physical theory. The goal of physical axiomatics is on the contrary to elucidate the peculiarities of *physical* theory in general, and the main characteristics of particular physical theories. And it employs formal tools which are created elsewhere: in pure mathematics. (For further contrasts see Salt, 1971.)

To work.

1. A FIRST EXERCISE IN AXIOMATISATION: NETWORK THEORY

We shall presently axiomatise the Kirchhoff-Helmholtz theory of electrical networks. We proceed right away to listing the presuppositions or background, the primitives or building blocks, and the axioms or postulates.

Formal background: ordinary logic (predicate calculus with identity), graph theory, and elementary analysis, as well as the set-theoretical, algebraic, arithmetical, and topological theories presupposed by analysis.

Philosophical background: semantics (the theory of meaning and truth) and the metaphysical presuppositions of scientific research (e.g. the independence and intelligibility of the external world).

Protophysical background: elementary systems theory, the elementary theory of universal time, and dimensional analysis.

Primitive base: T[time], G[graph], V[potential], e[e.m.f.], i[current], R[ohmic resistance], C[capacitance], L[self-inductance], and M[mutual inductance].

1. *Axioms on time*

(1a) T is an interval of the real line [*FA*].

(1b) Every member t of T represents an instant of time, and the relation $\leq$ that orders (partially) T represents the relation of being prior to or simultaneous with [*SA*].

2. *Axioms on network*

(2a) $\{G\}$ is a nonempty family of directed graphs [*FA*].

(2b) For every electrical network there exists a member G of $\{G\}$ that represents [models] the former in such a way that every terminal or junction is assigned a vertex of G and every element is assigned an edge of G [SA].

3. *Axioms on potential and current*

(3a) e, V and i are real valued functions of bounded variation on the set of ordered pairs $\langle$ edge, $t \rangle$ and they are continuous with respect to t [FA].

(3b) If n is an edge of a graph $G \in \{G\}$ representing an electrical net-work, then $e_n(t)$ represents the impressed voltage, $V_n(t)$ represents the electrical potential, and $i_n(t)$ represents the intensity of the electric current at the nth branch of the network represented by the nth edge of G [SA].

4. *Axioms on parameters*

(4a) R, C and L are real valued functions of bounded variation on $G \in \{G\}$ and M is a symmetric square matrix every element of which is a real valued function of bounded variation on $G \times G$ [FA].

(4b) If n and p are edges of a graph $G \in \{G\}$ representing an electrical network, then R_n represents the ohmic resistance, C_n represents the capacitance and L_n represents the self-induction of the nth branch of the network, while M_{np} represents the mutual inductance between the nth and the pth branches of it [SA].

5. *Axioms on laws*

If $G \in \{G\}$ represents a network in equilibrium (steady state), then:

(5a) At every vertex of G the sum of the current intensities along the branches represented by the edges meeting at the given vertex equals zero [PA].

(5b) For any loop of G, the sum of the potentials of the branches represented by that loop vanishes [PA].

(5c) For an arbitrary edge n between two vertices a and b of G,

$$L_n(di_n/dt) + R_n i_n + (1/C_n) \int dt \, i_n +$$
$$+ \sum M_{np}(di_p/dt) + e_n = V_n(a) - V_n(b) \; [PA]$$

Comments. (i) Our axiom system contains four more postulates than the usual three – namely the law statements (5a) through (5c). The function of our additional axioms 1 through 4 is to set the stage for the appearance of the law statements, which would make no sense otherwise. In other words, the preliminary axioms 1 to 4 specify the nature – but not the interrelations – of the nine primitive (undefined) concepts of the theory. This specification is both formal (mathematical) and factual (physical). The former is given by the axioms called *FA* (formal assumptions) while the content is sketched by the axioms called *SA* (semantical assumptions). In a heuristic presentation these additional assumptions are hinted at but not formulated explicitly. (ii) Sometimes a theory provides just a convenient (exact or suggestive) *language* – as is the case with information theory in genetics. If this is so then that theory is not really a presupposition of the given scientific theory, except perhaps heuristically. In our case, graph theory provides both a language and a body of theorems that facilicitate the search for and proof of statements about networks – whence graph theory is a genuine constituent of the background of network theory. (iii) The second members of every axiom group 1 through 4 contains the key semantical word 'represents'. Thus Axiom (2b) does not state that an electrical network *is* a directed graph but that it is *represented* or *modeled* by the latter. The reasons are these: (*a*) graphs are not things but ideas, and (*b*) any given graph can represent a whole class of equivalent real networks. (iv) Were it not for Axiom 4, the circuit parameters would be regarded as numbers. This postulate states that they are physical properties of networks. Since this is a phenomenological or black box theory, it does not tell us how R, C, L, and M come about: this is a task for mechanism theories such as Maxwell's, electrochemistry, and solid state theory. (v) The two Kirchhoff laws blend two different aspects of a network: the topological side and the physical one. This can be seen even better by adopting the matrix representation. In this representation, the currents and potentials for the various edges are collected in column matrices i and V operated on by the so-called vertex matrix A and the circuit matrix B respectively. In this formulation, the Kirchhoff laws are written: $Ai = 0$ and $BV = 0$, where A and B summarize the topological characteristics while i and V are the physical variables (Seshu and Reed, 1961).

2. Second Example: Classical Gravitation Theory

Let us now axiomatise the Newton-Poisson theory of gravitation. This will help us understand its relation to classical mechanics, with which it is often confused.

Formal background: logic and analysis (particularly potential theory), as well as the set-theoretical, algebraic, arithmetical, and topological presuppositions of analysis.

Philosophical background: the semantical and metaphysical presuppositions of scientific research.

Protophysical background: elementary systems theory, dimensional analysis, the theory of universal time, and physical Euclidean geometry.

Primitive base: M^3 [differentiable three-manifold], T [time], Σ [body], B [body representative], K [reference frame], Γ [field], U [potential], X [particle position], ρ [body density], $\mathbf{T}$ [mechanical stress], G [gravitational constant].

Axiom group 1: *space and time*

(1.1a) M^3 is a three-dimensional metric space with metric tensor g [*FA*].

(1.1b) M^3 represents ordinary space [*SA*].

(1.2a) T is an interval of the real line [*FA*].

(1.2b) Every member t of T represents an instant of time, and the relation $\leqslant$ that orders (partially) T represents the relation of being prior to or simultaneous with [*SA*].

Axiom group 2: *gravitational field*

(2.1a) Γ is a nonempty set [*FA*].

(2.1b) Every $\gamma \in \Gamma$ is a gravitational field [*SA*].

(2.2a) $\{U_\gamma\}$ is a nonempty family of scalar fields on M^3 [*FA*].

(2.2b) For every $\gamma \in \Gamma$ there is a $U_\gamma \in \{U_\gamma\}$ such that U_γ is a real valued function on $M^3 \times T$ [*FA*].

(2.2c) Every $U_\gamma \in \{U_\gamma\}$ and its first order derivatives are smooth on M^3 [*FA*].

(2.2d $-\nabla U_\gamma(x,t)$ represents the intensity of the gravitational field $\gamma \in \Gamma$ at $x \in M^3$ and $t \in T$ [*SA*].

(2.3) G is a positive real number [*FA*].

(2.4) For every $\gamma \in \Gamma$ and every $\sigma \in \Sigma$, at any point $x \in M^3$, at any instant $t \in T$ and in [relative to] any frame $k \in K$,

$$\nabla^2 U + 4\pi G\rho = 0 \quad [PA].$$

Axiom group 3: *body and frame*

(3.1a) Σ is a nonempty enumerable set disjoint from Γ [*FA*].

(3.1b) Every $\sigma \in \Sigma$ is a body [*SA*].

(3.2a) B is a nonempty family of point sets [*FA*].

(3.2b) Every $b \in B$ is a 3-dimensional differentiable manifold [*FA*].

(3.2c) For every $\sigma \in \Sigma$ there exists a $b \in B$ such that b represents [mirrors, models] σ pointwise [*SA*].

(3.3a) K is a nonempty enumerable set included in Σ [*FA*].

(3.3b) The distance between any two points in any $k \in K$ is constant [*PA*].

(3.3c) No $k \in K$ interacts with any $\sigma \in \Sigma$ that is not part of k [*PA*].

(3.3d) For every $k \in K$ there exists in M^3 a Cartesian system of orthogonal axes $e = \langle e_1, e_2, e_3 \rangle$ such that $e \cong k$ (i.e. e models or mirrors k) [*SA*].

(3.4a) $\{X\}$ is a nonempty family of real vector valued functions on $B \times K \times T$ [*FA*].

(3.4b) Every $X \in \{X\}$ is of bounded variation for any given $b \in B$ and $k \in K$ [*FA*].

(3.4c) If π is a particle of σ, and if $\beta \in b$ and $\beta \cong \pi$, then $X(\beta, k, t)$ represents the location of π relative to the frame k at the instant t [*SA*].

(3.5a) $\{\rho\}$ is a nonempty family of functions [*FA*].

(3.5b) Every $\rho \in \{\rho\}$ is a function from $B \times M^3 \times T$ to the set of nonnegative reals, Lebesgue integrable over any finite region of M^3 [*FA*].

(3.5c) If $b \cong \sigma$, then $\rho(b, x, t)$ represents the mass density of σ at x, t [*SA*].

(3.6a) $\{T\}$ is a nonempty family of functions [*FA*].

(3.6b) Every $T \in \{T\}$ is a real tensor valued function of valence $(2, 0)$ over $B \times K \times M^3 \times T$ [*FA*].

(3.6c) If $b \in B$ and $\sigma \in \Sigma$ and $\beta \in b$ and if π is a particle of σ, and furthermore $b \cong \sigma$ and $\beta \cong \pi$, then $T(\beta, k, x, t)$ represents the body stress at the particle π of σ [*SA*].

(3.7) For every $\gamma \in \Gamma$, every $b \cong \sigma$, every $\rho \in \{\rho\}$, every $X \in \{X\}$, every $T \in \{T\}$, every $x \in M^3$ and every $t \in T$, there exists at least one $k \in K$ such that

$$\rho \ddot{X} = - \rho \nabla U + \operatorname{div} T \quad [PA]$$

Derived concepts

Df. 1. *Total mass*:

$$b \stackrel{\triangle}{=} \sigma \Rightarrow M(\sigma, t) =_{df} \int_{X(b,t)} d^3x \, g^{1/2} \, \rho(b, x, t), \quad \text{with} \quad g =_{df} \text{Det} \, g_{ij}.$$

Df. 2. *Gravitational force density* on σ relative to $k \in K$:

$$f(\sigma, k) =_{df} - \rho \nabla U.$$

Df. 3. *Inertial frame*: Any reference frame such that the postulates 3 are satisfied is called an *inertial frame*.

Among the infinitely many consequences entailed by the preceding set of axioms, we shall mention only the following.

Theorem 1. The gravitational potential of a point particle of mass M is

$$U(r) = GM/r.$$

Proof. Set $\rho(r) = M\delta(r)/r^2$, where δ is Dirac's delta, in Axiom 2.4 and express ∇^2 in spherical coordinates.

Corollary. The gravitational force exerted on a particle of mass m by the field associated to a point particle of mass M is

$$F = GmM(X - r)/|X - r|^3.$$

Proof. By Thm. 1 and Df. 2.

Theorem 2. The equation of motion of a [spinless] particle of mass m in the field of a point particle of mass M is

$$\ddot{X} = GM(X - r)/|X - r|^3.$$

Proof. Replace Thm. 1 in Axiom 3.7, set $\mathbf{T} = 0$, and cancel ρ.

Corollary. Under the conditions of Thm. 2 and for nearly constant interparticle distances,

$$\ddot{X} = g = \text{const.,} \quad \text{with} \quad g =_{df} GM(X - r)/|X - r|^3.$$

The logical relations among the most commonly used physical assumptions and results of this theory are exhibited in the following tree:

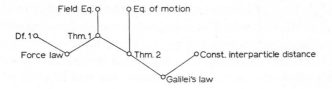

Comments. (i) The preceding axiom system contains only four physical assumptions: the ones concerning the rigidity and passivity of reference frames (Axioms (3.3b) and (3.3c) respectively), the field equation (Axiom (2.4)), and the equation of motion (Axiom (3.7)). The remaining 26 axioms are either mathematical or semantical assumptions. (ii) Even Axiom (2.3), concerning the scale G of the gravitational potential, is a mathematical assumption. On the other hand a statement concerning the dimensions of G may be regarded as a physical proposition, for it follows from the law statements in conjunction with dimensional analysis. Since it is a theorem, it need not be introduced in the axiomatic foundations of the theory. And a statement concerning the precise numerical value of G is a physical statement as well, but it would not be an assumption either, for it is entailed by the laws in conjunction with bits of empirical information (e.g., data concerning the length and period of a pendulum). (iii) The field equation is formally identical with the classical equation of the electrostatic field – which is often a source of puzzlement for beginning students. Were it not for the difference in the ponderomotive forces of the respective fields, we would be unable to discern the differences between these two fields. This is a reason for including the equations of motion in the theory. (iv) The elementary expositions of this theory are usually restricted to the most famous physical law, namely Theorem 2, which holds only for point particles. The textbook attempts to obtain the general equation of motion (3.7) by starting from a set of point particles is doomed to failure for obvious mathematical reasons. (v) The kinetic action of a gravitational field is mass-independent only in the special case when the stress has a vanishing divergence. This, then, is one of the restrictions under which a static homogeneous gravitational field is equivalent to an accelerated frame. Were it not that the value of div $\mathbf{T}$ is quite often negligible, and often neglected for want of information about $\mathbf{T}$, the equivalence principle (actually one of the two theorems that go by this name in general relativity: see Bunge, 1967a) might not have been discovered and it might not have helped build the relativistic theory of gravitation. (vi) According to the equations of motion (Axiom (3.7)), the kinetic effect of the inner stress, i.e. div $\mathbf{T}/\rho$, will, if negative, counteract and exceptionally even balance the ponderomotive action $- \nabla U$ of the field.

The two axiom systems we have expounded must suffice as a sample of physical axiomatics in the spirit of what was said in Chapter 7. For further examples in the same style the reader is invited to look elsewhere (Bunge, 1967a).

3. AXIOMATISATION TECHNIQUE

There are no techniques for constructing theories *ab initio*. Consequently no machine can be designed and programmed to construct theories from scratch – not even when feeding it an unlimited number of data. The building of theories is as creative, muddy and unruly a process as the writing of a poem or of a symphony (see Bunge, 1962a). On the other hand there are some rules of thumb that *help* to relativise and quantise classical theories provided these are not too complex and the relativiser (or the quantiser) realises the qualitative differences among such theories and the ambiguities arising in the transitions. There are also some heuristic rules for *reformulating* a physical theory in an axiomatic way – but the successful application of such rules presupposes an intimate acquaintance with the naive or intuitive formulations as well as with their applications in a number of examples. Hence the axiomatisation rules may not be fed to a machine either.

Once a physical theory has been produced and formulated in a reasonably clear way, it can be axiomatised by following these directions:

(i) *Review the main existing formulations of the theory and do not accept uncritically what passes for the best of them*: the chances are that it misses important hypotheses, or that it includes idle assumptions, or that it fails to meet the current standards of logical and mathematical rigor, or that its physical interpretation is too thin or even inconsistent.

(ii) *Collect all the main standard formulas actually employed by the practitioners of the field.* That is, gather all the most general statements of the theory and only those that are used in solving important typical problems. Pay more attention to what people do with theories rather than to what they say about them.

(iii) *Arrange the previous statements in order of generality*, starting with those (if any) that specify no particular model. These will be candidates for either central axioms or chief theorems of the axiomatic theory.

(iv) *Spot the main concepts in the preceding statements.* Some of them will prove to be the primitives of the theory.

(v) *Make a preliminary partition of the set of main concepts into primitive and defined.* Be sure to start off with the concepts that denote the physical system concerned: otherwise you may not know what you are talking about.

(vi) *Reformulate the key statements* (3rd step) in terms of the specific primitive candidates alone (5th step), using in addition as many logical and mathematical ideas as needed.

(vii) *Scan the preceding set of statements and try to derive the more specific from the more general statements.* If necessary to this end, add some further assumptions. Those statements that cannot be so derived are likely to be either strangers to the theory or components of a particular model of the subject concerned rather than ingredients of the general theory.

(viii) *Collect all the proving statements,* or premises, and set aside the proved ones. The former will belong to the axiom basis of the theory.

(ix) *Produce a revised list of primitives* by examining the main concepts occurring in the statements gathered in the 8th step. (Some new primitives may have crept in by way of the additional premises introduced in the 7th step.)

(x) *Lay down the mathematical and semantical conditions* the primitives must obey in order to satisfy the axiom candidates nominated in the 8th step.

(xi) *Collect all the candidates for postulates* produced by the 8th and 10th steps.

(xii) *List all the theories presupposed by the previous statements*: these will constitute the background of the given theory.

(xiii) *Collect the outcomes* of steps ix, xi and xii, i.e. list the presuppositions, the primitives, and the axioms of the theory. An – hardly *the* – *axiomatic foundation* of the theory will be ready.

(xiv) *Check* whether the foregoing includes or entails the standard formulas of the theory (step ii). If not, consider adding or deleting some axioms.

(xv) *Check* whether the axiom system entails any obviously false consequences. If it does, try to trace the sources (the derivations and/or the axioms) and modify them until the undesirables are removed. Replace them if necessary.

(xvi) *Check* the axiom system for consistency, primitive independence, axiom independence, and eventually other metamathematical properties as well – should you have any energy left.

The last step – the metamathematical analysis of an axiom system – is rarely undertaken. The reasons for this deficiency are clear. First, metamathematical tests are often hard to perform. Second, foundational workers are usually in a hurry to handle further theories. Third, one trusts his nose – even though it often scents wrongly. Nevertheless, a systematic exploration of the properties of axiom systems is called for and it would be at least as rewarding as the analogous investigation of mathematical theories, which is always bound to enhance their cogency and even their beauty.

It goes without saying that the preceding rules of procedure have got to be applied judiciously and imaginatively if they are to produce any valuable results at all. Not even the reconstruction of theories is a mechanical process: it takes some flair and experience to detect the key ideas in a theory and to strike a balance between pedantic rigor and utter sloppiness.

4. PROPERTIES OF A GOOD PHYSICAL AXIOM SYSTEM

Let us examine the properties an axiom system can have, and find out which ones are desirable in physics and why.

(i) *Formal consistency*: the axiom system should be free from contradictions. Otherwise it would entail every possible statement and consequently it could be used to "prove" anything whatever. That anything follows from a logical falsity is immediate: if A is false then $A \Rightarrow B$, with B arbitrary, is logically true and therefore immune to experience. Hence, by definition of entailment, A will entail B, whatever statement B may stand for.

Everyone agrees to the condition of formal consistency as the prime requirement of rationality and therefore as a condition every theory ought to satisfy. Nevertheless, the condition is often violated. Thus it is widely believed that a field theory can acquire a physical meaning only by resorting to the fiction of a passive test body, which moreover would supply an "operational definition" of the field strength. However, it is recognised that a test body that does not react on the field fails to satisfy the field equations and, moreover, is a perfect stranger in the case of a matter-free radiation field. It is also clear that the function of a test body is not to infuse meanings but, at best, to test a field theory. Even the latter is a fiction, for any actual field measuring instrument is far more complex than the mythical test body obediently moving along a force line. Moreover, field strengths (or the corresponding potentials) are not introduced by way of definitions but by way of axioms. Something similar happens with all other attempts to assign physical meanings in the spirit of operationalism: they introduce confusions between the referent of a theory and a way of testing it, and they shift the attention from the object or referent of the theory to more or less phoney (or else much too specialised) instruments that are allegedly described by the theory concerned – the truth being that an account of any real experimental set-up calls for the cooperation of a number of theories, as will be seen in Chapter 10.

(ii) *Deductive completeness*: the axiom system should contain (as axioms) or entail (as theorems) all the known law statements in the field that the theory is purported to cover, in particular the equations of motion and/or the field equations and/or the constitutive or state equations. Deductive completeness takes care of the desideratum of maximal truth value or degree of truth: indeed, the law statements in any field are the best available conceptualisations of the objective patterns the given discipline is concerned with. And in this context 'best' means "truest". Should any axiom system fail to cover a law statement in the given field, it would have to be enriched either by the addition of that law statement as an extra axiom or by strengthening some of the axioms so as to obtain that law statement.

Obvious as it sounds, this requirement of (weak) deductive completeness is hard to fulfil. But at least it must be recognised as a highlight of axiomatisation. This is not the case with the axiom systems for quantum mechanics produced by mathematicians: they often fail to include the general Schrödinger equation or an equivalent of it and consequently do not allow one to predict anything. An axiom system that does not concern physical systems – but deals with mathematical objects, or else with concrete nonphysical objects such as observations – and that does not contain any law statement, fails to qualify as a *physical* theory.

Notice that our requirement of weak deductive completeness concerns only *law* statements: it does not stipulate that the axiom system should assume or entail every statement in the field. Physical axiom systems should be deductively complete in a *weak* rather than in a strong sense, for if they were complete in the latter sense then no new premises could be adjoined to them and they would remain inapplicable and untestable. (In fact, given any statement s in a certain field, s is already a member of a complete theory T in that field, so that s cannot be added to T. The other possibility would be to adjoin to T the denial of s, but this would introduce a contradiction: the theory T' enriched with the statement not-s would be inconsistent. In short, a complete theory in the strong sense cannot be enriched except by rendering it inconsistent. Equivalently: only incomplete theories can be adjoined further premises without risking contradiction.)

We want then our scientific theories to be *in*complete so that they can be enriched with non-law statements such as subsidiary hypotheses and data – e.g. the assumption of constant interparticle distance that was added in Section 2 to the axioms of the classical theory of gravitation in order to obtain Galilei's law of falling bodies. Otherwise, i.e. if our theories were closed to external (but congenial) premises, they would be inapplicable and untestable, for every application and every test requires the adjunction of statements (e.g. initial conditions, special values of functions, etc.) that are too specific to be included among the axioms. A complete theory would be either an ivory tower or an example of a theory, i.e. an application or theoretical model – and in either case it would be incapable of handling new cases. In short, we should attempt to axiomatise only the core of a theory.

At any rate complete theories are hard to come by, and incompleteness in the strong sense is welcome not only in physics but sometimes also in mathematics, and this for the same reason: because it affords the possibility of adjoining assumptions other than the axioms and so obtaining more specific theories. Thus one of the reasons for the versatility of general group theory is that it can be enriched with any number of assumptions – e.g. commutativity (to get abelian groups) and the condition that the basic set consists of a fixed number of elements. (Incomplete theories are sometimes said not to be axiomatisable, but this is wrong if they happen to be based on a definite set of axioms. What may not be fully axiomatisable is the whole field such incomplete theories purport to cover.)

Now, to implement the goal of deductive completeness in the weak sense (exhaustive coverage of the laws in a field), we must build strong axioms systems. That is, if we want a rich theory we must pick strong axioms. And if we want strong axioms we must use strong basic (undefined) concepts. A strong concept is one that subsumes may other concepts, just as a strong axiom is one that has plenty of logical consequences. We should therefore abstain from choosing singular statements (propositions concerning specific individuals) and, in general, we should try to discard idiosyncrasies when building axiom systems. Specifying details would be as foolish as legislating the diameter of pipelines: such as specification must be left to the applications. Even general but derivable statements must be discarded as candidates for the axiom status. Thus we should not postulate averages, since distributions will give us not just averages but any number of statistical moments as well. In short, we must favor logical *strength*, for the stronger an idea the richer its content. Let experiment clip our wings, but let us grow wings to begin with.

(iii) *Primitive completeness*: axioms other than the physical assumptions should lay down necessary and sufficient conditions for every one of the basic (undefined) concepts of the theory, so that these concepts make both mathematical and physical sense. Moreover, every such axiom should make sense by itself, so that it may be replaced or even negated when looking for improved theories or for independence proofs. This calls for a minimum of complexity. Thus the statement ". is a binary associative operation on the set S" should not be split into "S is a set" and ". is a binary associative operation," for it would be impossible to find a model (true interpretation) in which the one holds while the other does not.

The specification of the mathematical status (set, relation, function, etc.) of every primitive is a mathematical task that can be carried out to any degree of refinement allowed by the state of mathematics. On the other hand the task of assigning a physical meaning to a symbol is seldom performed in a thoroughly satisfactory manner, and this for both technical and philosophical reasons. The technical difficulty boils down to this: while in mathematics a theory is normally interpreted, if at all, within some other theory (e.g. the elements of a group as numbers), the interpretation of a physical symbol consists in assigning it an *extratheoretical* object – either a physical entity (e.g. a dielectric) or a physical property (e.g. the dielectric power). And such a physical correlate or referent of the symbol concerned is known in part through the very theories. Consequently the assignment of physical meaning is not done term by term and in a complete way: surely one must not forget to state the semantical assumptions, for they do trace at least a semantical profile of the primitives, but one must not think that they will endow the symbols with a clear-cut and full meaning. In short, physical meanings are assigned by whole theories and even so only in outline.

As to the philosophical obstacle to the performance of this task, it consists in the survival of a philosophy so distrustful of theories that it demands the reduction of every theoretical term to a complex of laboratory operations, rather than allowing for the theoretical explanation of the latter. Thus some physicists, wishing to endow general relativity with a physical content, and confusing meaning with testability, will cram the whole universe with rulers and clocks handled by ubiquitous observers. By so doing they overlook that such a profusion of measuring instruments and observers would distort the field out of recognition, and they forget that the addition of imaginary elements will not render the theory more realistic. If a theory is to be physical, it must be interpreted in physical terms: not in terms of human operations but in such a way that the interpretation assumptions assign the (basic) symbols supposedly objective referents, and in such a way that these assumptions (which may prove false) do not contradict the remaining assumptions of the axiom system.

(iv) *Primitive independence*: the basic concepts of an axiom system should be mutually independent, i.e. they should not be interdefinable. (If any of them were in fact definable in terms of other basic concepts, then it would not be a primitive concept.) The importance of this property lies not so much in economy as in that it focuses our attention on the logically basic units and it prevents circular moves such as the attempt to define mass as an acceleration ratio (on the basis of Newton's law of motion) and then force as the product of mass by acceleration.

(v) *Postulate independence*: ideally, the various axioms of a theory should not be interdeducible. (If one of them were derivable from some other axioms of the theory then it would be a theorem of it.) This condition is important not so much because of its alleged ink-saving property – which is illusory anyway – but because it facilitates the alteration of the theory in the light of experimental or theoretical developments. For, if the axioms responsible for a false consequence can be spotted and removed, then the others may be kept. In short, independence facilitates theoretical progress.

5. UNDESIRABLES

Thus far, five desirable characteristics of a physical axiom system have been pointed out. Let us mention a few undesirable ones. One such characteristic – namely, *completeness* in the strong sense – was mentioned before. A related property is the one of *categoricity*, or rather rigidity or inflexibility. A categorical theory is one such that any two models (true interpretations) of its underlying abstract formalism are isomorphic (structurally identical). Now a necessary condition for theory isomorphism is that the corresponding sets be similar, i.e. that there be a one-to-one correspondence between them. But we do not want such a rigidity in physics for, even if two theories do have formally identical basic formulas (e.g. wave equations), they may refer to entirely different kinds of physical systems, these kinds being conceptualised as sets that need not be similar.

How about *simplicity*, in particular formal simplicity or economy of form? It used to be claimed that every theory should have maximal formal simplicity, in the sense that it should have the least possible number of primitives and axioms. But this requirement is bound to be crippling unless the qualification is added, that the minimisation of basic ideas (primitives and axioms) must be consistent with the desirata of both weak deductive completeness and primitive completeness. For otherwise the simplification may go as far as to produce a false theory. In any case desiderata ought to be justified, and no reasonable justification is ever given for formal simplicity, save that it is handy.

Nevertheless there is one reason, namely that simplicity decreases the chances of error and, in particular, of hidden inconsistency. Thus if we can deduce the basic laws of a field from a single variational principle (accompanied by its mathematical and semantical setting), the consistency problem will hardly arise. (Or, rather, the burden of consistency proofs will be placed on the shoulder of the mathematician.) But if so, simplicity is a means not an end in itself. As for the other kinds of simplicity – semantical, epistemological, methodological, and pragmatical (Bunge, 1963) – they may occasionally be of value as long as they do not conflict with higher desiderata, mainly consistency and weak deductive completeness, which involves maximal truth. After all, the goal of scientific research is not to gratify philosophical prejudices such as simplicism and the hatred of theories, but to find out how things really are, even if they persist in their bad habit of being usually far more complex than we had presumed originally.

At any rate the number of primitive concepts of a theory cannot be diminished at will, under penalty of impoverishing the theory to the point of uselessness. The bare minimal primitive base of a qualitative physical theory is made up of two concepts: a reference class (the set of physical entities the theory is about) and another item, which could be a relation or an operation on that set – e.g. the physical addition or juxtaposition of two arbitrary elements of that set. Otherwise the theory could not contain a law statement, e.g. the commutativity of physical addition. A quantitative theory requires at least three primitives: the reference class and two other items, which could be a further set and a numerical function on the cartesian product of those two sets. Thus, about the simplest quantitative formula one can think of is of the form: $dP/dt = 0$, where P stands for a property and is a real valued function on $\Sigma \times T$, Σ being the set of referents and T the set of instants.

Now, a theory with two primitives needs at least five axioms: one physical assumption relating the two primitives, and two nonphysical (mathematical or semantical) assumptions for each primitive: one mathematical and one semantical axiom. In general, N primitives call for a minimum of $2N + 1$ axioms. This is a bare minimum: normally it is impossible to compress all the mathematical properties of a concept into a single statement – unless one resorts to the trick of conjoining several axioms. (This trick was resorted to in Section 1: Axiom (3a) is actually the conjunction of three axioms.) No wonder, then, that even an elementary theory such as Newtonian particle mechanics, which has eight independent basic specific concepts, should contain more than two dozen axioms when spelt out (Bunge, 1967a).

6. ADVANTAGES OF AXIOMATICS

There are at least ten good reasons for valuing the axiomatic approach to physical theory:

(i) *The presuppositions are recognised and kept under control.* In the kind of axiomatics advocated in this paper, the background of a theory – both its formal and nonformal presuppositions – is exhibited in the first place, so that it can be kept in mind for eventual criticism and correction. Example: it is sometimes claimed that quantum mechanics presupposes classical mechanics. On the other hand it is known that these two theories are mutually inconsistent. The mistake could have been avoided upon axiomatising quantum mechanics: when this is done no such dependence on classical mechanics emerges.

(ii) *The referent of the theory is not lost sight of.* Unless formulated in detail, a scientific statement may seem to concern no real entity whatever, or else it lends itself to arbitrary interpretations. But if the axioms state explicitly what *all* the arguments (or the indices) of the functions occurring in the statements are, no such mistakes are likely to be made. Example: it is often held that quantum mechanics is not about autonomously existing physical systems but about measurements, or about unanalysable object-apparatus-subject blocks, or about our knowledge, or even about propositions. Thus it will be stated dogmatically that every hamiltonian one can think of represents an energy measurement – even in cases when it is apparent that the hamiltonian is about a free system and even though it never supplies any information as to how to perform such a measurement. In an axiomatic formulation no such groundless statements are allowed: one knows what one is talking about.

(iii) *Meanings are assigned systematically, consistently, and in a literal way* rather than erratically, inconsistently, and metaphorically. An open context allows for arbitrary and analogical assignments of meaning and therefore harbours ambiguity and invites inconsistency. Such risks are minimised in an axiomatic context provided the semantical axioms are supplied. Example: in the usual textbooks on quantum mechanics 'ΔX' is given a variety of mutually incompatible interpretations: mean standard deviation, subjective uncertainty, measurement error, width of a wave packet, etc. Moreover few of these interpretations are consistent with the interpretation assigned the remaining symbols of the theory; for example, there can be no question of subjective uncertainty unless a subject is explicitly introduced. An axiomatic formulation of quantum mechanics, on the other hand, will commit itself to a single definite interpretation of 'ΔX' and, what is more, to an interpretation consistent with the one assigned to 'X' and the other symbols, rather than to an adventitious interpretation (see Chapter 5.)

(iv) *Further theorems can be discovered.* If the initial assumptions of the theory are laid down explicitly and are couched in the best available mathematical language, the number of theorems is bound to increase – i.e. the known part of the theory is bound to grow. Example: by shaping up and axiomatising several chapters of classical physics, a number of new theorems have been derived in recent years. (See e.g., Truesdell, 1966, 1969.)

(v) *Invalid proofs are kept to a minimum.* In an open context, to prove a theorem one is tempted and even authorised to resort to any premises that may be of help. This cross fertilisation leads sometimes to new valuable ideas, and at other times to absurdities. It must be watched carefully in physics, which is a set of theories that are not all mutually consistent. The neglect of boundaries becomes dangerous, whether in physics or in mathematics, when it comes to proving metastatements, in particular statements concerning a whole theory, for in this case the theory concerned must be clearly "defined," i.e. it must be formulated axiomatically.

Example: all the reasonings involving "virtual" processes in atomic, nuclear and particle physics are invalid because they are supposedly justified by the so called fourth indeterminacy relation, $\Delta E \cdot \Delta t \geq \hbar/2$, which is meaningless in quantum mechanics as time is a scatter free variable in it (recall Chapter 2, Section 2). Consequently the meson theory of nuclear forces is unsound in that it assumes that the mechanism of strong interactions is the exchange of pions, construed as the reversible transformation of a proton into a neutron and a positive pion. These processes are impossible because they would violate energy conservation: indeed, the nucleon masses are nearly the same and the pion mass is about 140 MeV. However, these exchanges and the transfer pion are regarded as the very source of nuclear stability. They are called *virtual* (rather than *fictitious*) and are justified by recourse to the 4th indeterminacy inequality, which is itself unjustified. The justification is this: on setting $\Delta E = 140$ MeV (which is queer because no scatter around an average is involved) one gets $\Delta t > 10^{-24}$ sec. And this is too short a period for the violation to be observable, hence real in an operational sense. The whole construction crumbles down upon realising that it rests on the meaningless 4th indeterminacy relation.

(vi) *Irrelevant proofs are spared.* In an open context one may be tempted to seek proofs that are unnecessary or do not even make sense in the given context. Example: the assertion that there are no hidden variables is, strictly speaking, meaningless. What is meaningful is the relativised statement that there are no hidden variables in *standard* quantum mechanics, for this statement can be checked (and proved) by scanning the basic dynamical variables of the theory – which examination requires the axiomatisation of the latter. Part of the controversy over hidden variables might have been averted if genuine axiomatisations of quantum mechanics had been available. And the whole current discussion (see Bastin, 1971) on the possibility of going beyond quantum mechanics – either in the same direction, or else by reinforcing the stochastic ingredients, or finally by weakening them – would profit from such an explicit and cogent formulation, for then one would know exactly what is to be overcome and what kinds of new variables ought to be introduced.

(vii) *Utopian rationalism is avoided.* The naive rationalist wishes to define every concept and prove every statement: he does not realise that this will lead him either to turn in circles or to an infinite regress. Genuine rationality calls for the acceptance, at least *pro tempore*, of a set of undefined concepts and unproved statements, for they will enable us to derive and so justify all the rest. Surely, that acceptance is not a matter of faith: it must be justified. The justification for introducing a primitive concept is that it plays a role in a theory; and the justification for proposing an axiom is that it (usually in cooperation with other premises) entails theorems that explain or predict something. Example : the habit of starting an exposition with a list of definitions betrays utopian rationalism and so does many an attempt to deduce quantum mechanics either from classical physics or from purely mathematical theories.

(viii) *Heuristic insights are gained.* An axiomatic theory, by exhibiting all its assumptions, tempts the bold scientist to suspend some of them, with replacement or without it, just to see what "happens", i.e. how the set of consequences is affected. If a postulate is deleted, some theorems will be lost; if it is replaced by a different assumption, some theorems will change. In either case a new theory will have been produced. Example: non-Euclidean geometries were built in this way.

(ix) *Analysis is facilitated.* Usually, the analysis of physical ideas proceeds in no fixed context. Not even discussions about definability are normally carried out in an axiomatic context. In this way nothing can be settled, for it is perfectly possible for a concept to be undefined in one system and defined in another. Likewise a hypothesis may be assumed in one theory and derived in another. Open context analysis is necessarily inaccurate and incomplete: it neglects basic ideas and introduces irrelevant notions, and in any case it cannot exhibit the exact form and the meaning of a symbol, for to do this is precisely to produce an axiom system, however minute. Example: all the nonsense about defining the time concept in terms of irreversible processes might have been spared either by building an axiomatic theory of time or by axiomatising at least one theory of irreversible processes. The former move would have shown that a universal concept of time, one applicable in every chapter of physics, must not be tied to any special process. And the second move would have shown that some time concept must be available before the very equations for an irreversible process can be written out. (See Bunge, 1970e.)

(x) *The tampering with single formulas out of context is discouraged.* Suppose, or rather recall, that someone claimed that special relativity countenances the hypothesis that there are superluminal particles (tachyons) on the ground that this conjecture does not violate the definition of momentum in relativistic mechanics. An obvious rejoinder, from the point of view of axiomatics, would be this. Tachyons and other strange things may well exist in reality but, if they do, then they obey some theory other than relativistic mechanics, which involves the Lorentz formulas that tachyons disobey (Mariwalla, 1969), and has no use for imaginary trajectories. (If $v > c$ then there are no real trajectories for real forces unless the imaginary tachyon be assigned an imaginary mass – which is physically meaningless.) Something similar happens with the conjecture that there are particles with imaginary electric charge: it is true that they might interact through a real Coulomb force, but on the standard electromagnetic theory they could not interact with normally charged particles and they would be incapable of emitting electromagnetic waves. To generalise: while any given formula may be modified *ad libitum* when taken in isolation, an entire theory cannot be tampered with so easily, as it is a system in which all components hang together. This is why we are justified in assigning whole theories, or hypothetico-deductive systems, a stronger credibility than stray conjectures.

(xi) *Numerological acrobatics is avoided.* A purely formal game with physical constants and other numbers can be made to yield sets of numbers that look physically significant. Such Pythagorean games were popular in the 1930's and are bound to be played again unless we get used to think in terms of whole theories. Indeed, the requirement of axiomatisation exhibits the lack of significance of such games by showing (a) that they involve hardly any law statements and (b) that they fail to indicate clearly the referents of the symbols involved. (For the triviality of numerology see Chapter 3.)

(xii) *Metamathematical tests are made possible.* Unless a theory has been axiomatised one cannot be sure whether it has any of the metamathematical properties (e.g. consistency) one claims for it. (Axiomatisation is necessary but not sufficient. Even in mathematics one often gets, at most, proofs of *relative* consistency: thus Euclidean geometry can be shown to be consistent provided the consistency of the real number system is assumed.) Example: the existing proofs of the equivalence (isomorphism) of matrix mechanics and wave mechanics are heuristic rather than rigorous, and this for two reasons. First, the very definition of isomorphism has to be constructed *ad hoc* for every kind of theory: thus the definition for a primitive base consisting of a set and a relation differs from the one adequate to a base consisting of two sets. Second, no axiomatisation of quantum mechanics in either "picture" (formulation) was available at the time the proofs were given. Consequently, it is possible to doubt – as Dirac does it – whether the two formulations are actually equivalent. Something similar holds for Feynman's path integral formulation in relation to the standard formulations.

(xiii) *Recall is greatly facilitated.* Experimental psychologists have shown that a well-organised body of knowledge is far easier to recall than a set of items with no apparent ties. Indeed, recent experimental work has shown that "our memories are limited by the amount of units or symbols we must master, and not by the amount of information that these symbols represent. Thus it is helpful to organise material intelligently before we try to memorise it. The process of organisation enables us to package the same total amount of information into far fewer symbols, and so eases the task of remembering" (Miller, 1967, p. 12). Since our storing capacity is rather poor anyhow, we should derive psychological benefit from keeping in mind just the central axioms and a few typical theorems of a theory rather than a motley assembly of statements. The pedagogic potentialities of the axiomatic approach will be further examined in Section 8.

Let us now turn to the usual complaints over axiomatics.

7. STANDARD OBJECTIONS TO AXIOMATICS

The main standard objections to the axiomatic approach seem to be the following.

Objection 1: Axiomatics fails to portray the actual process of theory construction. Therefore it does not teach us how to construct theories. *Rejoinder*: True but irrelevant. The disclosure of actual research processes is the concern of methodologists, psychologists, historians of science and biographers. (Recall Chapter 7, Section 1.) One cannot attain systematicity and historicity at the same time, for the two are poles apart: either one works on a reasonably neat formulation of a theory, or one gives a good coverage of the story of its conception – a zig-zagging process that is full of more or less obscure motivations, as well as of irrelevancies, inconsistencies, and false moves. It is not a shortcoming of the axiomatic method but rather a virtue of it that it yields finished products of public property, the use of which does not require a prior acquaintance with the biographies of the scientists involved in motivating, building, applying and testing the theory.

Objection 2: Axiomatisation is rehashing rather than original work. It is a job for textbook writers, not for original investigators. *Rejoinder*: Mathematicians do not share this belief. They regard Euclid's axiomatisation of geometry, the Dedekind-Peano axiomatisation of arithmetic, Kolmogoroff's axiomatisation of probability, and Bourbaki's numerous axiom systems, as original works. They believe these axiomatisations (a) discovered the essential ideas and the logical relations in a theory, (b) clarified and purified the ideas involved, and so (c) facilitated further developments. Thus before Kolmogoroff's work on probability and the measure-theoretic foundation of it, people tended to focus on Bernoullian sequences and the statements concerning them, such as the law of large numbers and the principle (now a theorem) that no systematic deviations from probability should be expected for such (very particular) sequences. While these are important theorems, they are not logically fundamental and, because they concern a particular case (Bernoulli sequences), if one focuses attention on them (as von Mises did) one gets a wrong idea of the generality of probability theory and of its logical structure. If mathematicians value axiomatics, why should physicists despise it? (By the way, what is wrong with writing textbooks? Why should a good textbook rate less than a bad paper?)

Objection 3: Axiomatics is sterile. New laws are discovered by heuristic procedures, not by an axiomatic reconstruction of what is known. And new problems are solved by applying and enlarging existing theories rather than by reshuffling them. *Rejoinder*: Largely true but not quite. Indeed, (*a*) axiomatisation itself is a novelty and it exhibits previously unknown or hidden traits, and (*b*) axiomatisation does have some heuristic power, as it facilitates the expansion and criticism of available theories as well as their replacement by improved theories. (Recall Section 6.) Furthermore, the objection is irrelevant, for the main goal of axiomatics is not to find new laws but to house them properly. Nor is it to solve new problems *within* a theory but rather to answer questions *about* a theory. Indeed, a secondary aim of axiomatics is to contribute to our knowledge about theories, for it is only after axiomatisation that their structure and content can be clearly discerned, and that they can be compared with rival theories and thus evaluated. (See Chapter 9.) Even if the axiomatisation of a theory does not render it more powerful – i.e. does not give it a greater coverage and depth – it will render it more exact and perspicuous, so that its merits and shortcomings can be better assessed, and consequently the debates over the theory can be fruitful rather than irritating. Also, the philosophy attached to the theory can best be disclosed upon axiomatisation.

Objection 4: Axiomatisation is a strait-jacket: it prevents the further development of the theory. *Rejoinder*: Quite on the contrary, the explicit display of assumptions makes it easier to obtain further consequences as well as to criticise and evaluate a theory. As long as a theory is shrouded in vagueness and confusion it can lend itself to endless and fruitless apology and controversy. If a theory is worthless, then its axiomatisation will show this in an unequivocal way; if it contains valuable germs, these can best be grown and protected from pernicious partners upon explicitly and orderly displaying all the components, the good and the bad ones.

Objection 5: Axiomatics is authoritarian: by calling a mere hypothesis an *axiom*, we feel awed: our critical powers founder and we are led to believe what we should doubt. *Rejoinder*: This attitude betrays a magical fear of words and a ignorance of both the etymology and the present day meaning of the word 'axiom'. The Greek word ἀξίωμα, just like the

Latin *postulatus*, mean "request." So do 'axiom' and 'postulate' mean to us, after having freed ourselves of the philosophical tenet that axioms should be self-evident and beyond criticism. By writing down a formula and calling it an axiom all we do is to *ask* everyone concerned to examine it rather than to believe it: to consider it, not to accept it. We do not attempt to bully anyone into admitting uncritically the initial hypothesis we have dignified with the title of axiom. By writing our initial assumptions in an explicit fashion we prepare the ground for a possible fruitful argument – with ourselves in the first place, as the very first thing we are to do with an axiom system is to find out and assay what it entails. It is only by refusing to exhibit all our assumptions (hypotheses) that we may hope to pass counterfeit bills. Axiomatics invites scrutiny, criticism, dialogue.

Objection 6: Axiomatics is limited anyway, so why bother? Indeed, every powerful theory is incomplete in the first place. (Recall Chapter 7.) Secondly, axiomatics does not tell us how to verify a theory. *Rejoinder*: Both objections are correct. Nevertheless they are wide of the mark: the limitations of a tool do not render it useless. As to the inherent incompleteness, in the metamathematical sense, of every rich theory: (a) mathematicians face the same limitation, which does not deter them from using the axiomatic format; (b) we may well settle for a weaker sense of 'completeness': we may say that a physical theory is deductively complete in a weak sense iff it contains all the standard theorems in the field. (Recall Section 4, point 2.) And concerning the inability of an axiom system to instruct us as to how to use and test it, this is an advantage rather than a shortcoming: it is a sign of generality. If a theory, whether axiomatic or informal, does not specify the circumstances under which it can be applied or put to the test, then it can be enriched with further tentative premises concerning particular kinds of thing and special circumstances. In this sense classical electrodynamics is complete provided its domain is restricted to macroevents; on the other hand classical thermodynamics is radically incomplete because it does not apply to the ever-present nonequilibrium processes. In any case, while axiomatics does fail to be perfect (in the sense of all-inclusiveness), it is the best possible organisation a theory can be given. So why fight it?

Objection 7: The basic concepts of an axiom system, by remaining undefined, stay unanalysed and therefore obscure. *Rejoinder*: True in

Aristotle's time, false nowadays. The objection rests on an outdated theory of definition. We have learned that definition is only one kind of analysis and elucidation – and one that cannot always be performed under penalty of circularity. Another, more exhaustive kind of analysis and elucidation, is the one performed by the axioms characterising the form and content of the undefined ideas.

  Objection 8: Axiomatisation, being a purely formal procedure, is incapable of capturing the factual meaning of a theory. *Rejoinder*: true of *formal* axiomatics, nearly false of *physical* axiomatics. (Recall Chapter 7, Sections 8 and 10.) While the former ignores physical content altogether, physical axiomatics systematises the intended interpretation of the formalism, thus adding a semantical precision absent from formal axiomatics as well as from the informal or heuristic presentations. In informal theorising, meanings are vaguely indicated by the context: one therefore speaks of the *intended* meaning of the symbols involved. Moreover, in open contexts there is no guarantee of consistency – formal or semantical. Only physical axiomatics makes a firm semantical commitment at the axiom level and carries it through all the way down to the theorems. Even so, it can only trace a semantical profile of a theory: unlike form, content remains always somewhat hazy.

  In conclusion, the main standard objections to axiomatics seem to derive from insufficient acquaintance with it.

8. THE PLACE OF AXIOMATICS IN TEACHING

Axiomatics is not meant for the beginner: before any order can be brought to a subject, the latter should have been grasped in an informal or heuristic way. A premature exposure to axiomatics may result in incomprehension or in boredom. Witness the teaching of Euclidean geometry for centuries before the discovery that children are not small-scale adults. Hilbert himself, the champion of axiomatics in all fields of learning, was aware of the pedagogical and psychological limitations of the axiomatic approach and advised a judicious compromise between it and the heuristic or genetic approach – in teaching, that is. He went further, by co-authoring a textbook on intuitive geometry.

  When should a first exposure to axiomatics occur? As early as possible if lots of mistakes, obscurities and overlaps are to be avoided, and if a grasp of fundamentals is preferred to the transient memorising of masses of unrelated items. But when is it possible to do so? Clearly, the width of the axiomatic threshold depends on the subject. Modern algebra can be taught and presumably must be taught the axiomatic way from the beginning, even at the high-school level (Suppes 1966). But physics is much more complex than algebra – even more so than the calculus, which cannot be taught axiomatically to students lacking a minimum of mathematical sophistication and an ability and liking for abstract thinking. It seems obvious that elementary physics should continue to be taught the heuristic way, if only because the understanding of a physical axiom system requires mastering certain logical and mathematical ideas that are acquired later on. But the *instructors* should be aware of physical axiomatics, so that they will abstain from repeating scientific mistakes (like equating mass and body, or energy and radiation), logical mistakes (like trying to define everything, or to prove assumptions by showing that some of their consequences are true to fact), or philosophical mistakes (like confusing concept and proposition, or law and rule, or like claiming that all theories can be inferred from experimental data).

The axiomatic approach should be tried at the graduate level. But even here it might be mistaken to discard the heuristic approach altogether. The writer has tried with success the following compromise between heuristics and axiomatics: *three-quarters heuristics and one-quarter axiomatics.* The first three-quarters of the time may be devoted, as usual, to an informal (but not necessarily wrong and messy) presentation of the main assumptions and the main theorems, with an abundant sprinkle of exercises and problems. By the end of this period the inquisitive student will have encountered so much material and in such a disorderly and lacunar way, that he will look forward to a cogent and concise presentation of the foundations of the theory. Having acquired a good stock of more or less isolated formulas, he will be ready for an axiomatic presentation of the whole, which may take as little as a couple of weeks. This exposure to axiomatics will give the student a chance to rehearse the material, to organise it better, to go deeper into it, and to analyse it critically. For an axiomatic presentation unaccompanied by a critical analysis is one more piece of dogma. And this – the critical analysis of an axiom system – is of course the opportunity to add a pinch of methodology, another of philosophy, and an icing of history. Since all this is done anyway, better do it explicitly and in the glaring light of an axiom system than surreptitiously and in the *chiaroscuro* of heuristics.

In short, physical theories should be taught axiomatically once their fundamentals have been grasped.

9. CLOSING REMARKS

To axiomatise is just to maximise explicitness and articulateness. Those who care little for either need not bother about axiomatics, but those who do care will not settle for less, or they will at least tolerate those who try to organise the rather untidy products of original research.

No one should rate axiomatics in science higher than the creation of new powerful theories. Yet a suitable axiomatisation of a good but controversial theory is certainly no less valuable than the fashioning of a bad and ignored theory. Axiomatisation does not replace the creation of theories and does not compete with it but, on the contrary, consummates that creative process. Like every other refinement, axiomatisation is convenient, nay optimal, rather than indispensable for everyday purposes.

However, just as there are occasions which call for biscuits rather than bread, so there are in science certain crossroads where ordering is more valuable than jamming. If the problem is to clarify theoretical and methodological issues, to analyse and evaluate theories, and to assess rival programmes of theory construction rather than to work out and apply existing theories, then axiomatics ceases to be a refinement to become a prime necessity. Indeed, only clearly and fully formulated theories can be given a fair trial.

Also, axiomatics can help the maturation of physical science rather than its mere growth in bulk. Indeed, axiomatics enhances cogency and clarity – hence exposure to analysis and criticism – which, together with depth or boldness, constitute maturity as distinct from mere size (Bunge, 1968a). Finally, axiomatics can help us meet the information explosion, or rather deluge. For, if we cannot keep up with details, we can at least keep up with the development of fundamental research in a given field: foundation problems are always "in" and final solutions to them are seldom to be expected.

CHAPTER 9

THE NETWORK OF THEORIES*

Every single physical theory can be neatly articulated, i.e., axiomatised. How about the whole lot of theories? Would it be possible to organise them into a single gigantic system with a fundamental or comprehensive theory at the basis and the various regional theories as so many logical consequences of that basis? Such an organisation has been dreamed of several times. A century ago mechanics was regarded as the foundation of the whole edifice of physics; later on the hope shifted to electro-dynamics; more recently, to general relativity or to quantum mechanics. But so far no thoroughly unified (or rather unifying) physical theory has been found that accounts for the whole of physical reality and contains every particular theory. What has been found is (a) that the number of theories keeps increasing, (b) that some theories, formerly be-lieved to be autonomous, prove to be subtheories of others, and (c) that whole classes of theories can be formally subsumed under comprehensive formalisms, such as the lagrangian one which, alas, has hardly any physi-cal meaning. Obviously none of these partial successes meets the require-ment of thorough mathematical and physical unification. So, we remain practical pluralists even though some of us may dream of a single super-theory. And, understandably enough, most of us are more anxious to evolve theories accounting for uncharted territories, like high energy physics, than to unify the existing, obviously insufficient theories.

That the existing physical theories cannot be brought together under a single comprehensive theory without changing considerably in the pro-cess, should be clear: our present theories are not thoroughly consistent with one another. And we cannot afford to dispense with any of the great theories even knowing that they have incurable shortcomings. For exam-ple we need classical mechanics, which is partially inconsistent with both electrodynamics and quantum mechanics, if only to devise tests for the latter theories. A unifying theory cannot therefore be a mere fusion of the available theories: it must be radically new. Whether such a dream will come true, is impossible to forecast. In any case it is a problem for physicists not for workers in the foundations and philosophy of science, as it consists in actually building a physical supertheory (not just a mathematical one) rather than in reorganising or analysing any existing theories.

Whether or not a unifying supertheory is possible, we should map out the world of actual physical theories: we should be able to exhibit their mutual relations, much as mathematicians can do it with most of their theories. Unfortunately this too is an open problem: although anyone can draw pretty diagrams exhibiting the supposed relations among the various physical theories, there are hardly any *proofs* that such relations do in fact hold. And there are no proofs because it is not generally realised that such metatheoretical statements require a proof. And, even if the need is acknowledged, we do not quite know how to proceed with such a proof: the tools are there but we do not handle them deftly.

This chapter has three aims. One is to review some of the tools required to explore the relations among physical theories. Another is to exhibit the richness of the intertheory relations, hoping that this may serve as a reminder of the complexity and backward state of the problem – hence as a stimulus for a deep and unified approach to it. The third goal will be to show that many problems about intertheory relations, which are usually regarded by both scientists and philosophers as solved, have hardly been posed in a correct way.

1. PRESENT STATE OF THE PROBLEM

1.1. *Three Parallel Studies*

As with other metascientific problems, scientists as well as philosophers have contributed to the literature on the relations among theories. And, as usual, the two groups have done their best to ignore each other. In this case they have also managed to ignore a third group, which happens to be the most articulate of all: namely the logicians and mathematicians who have created the calculus of theories, model theory and category theory, and have studied the formal relations among hypothetico-deductive systems. The unfortunate result of this lack of communication among the three groups is that we have three disjoint sets of studies. It is an urgent task of metascientists to intertwine these three separate threads with a view to producing a unified picture of intertheory relations.

The scientists concerned with this problem have dealt almost exclusively with one type of intertheory relation, namely the one that obtains when two theories with roughly the same intended referent (recall Chapter 4) have different extensions or coverages, and when certain characteristic parameters of the one approach a limit – e.g. when the velocity of light in vacuo goes to infinity, or when Planck's constant is set equal to zero. While this is an interesting and important case, it does not exhaust the relations among theories. Moreover, it has yet to be treated in a general and rigorous way.

The philosophers, who are expected to examine all the sides of a problem, have concentrated on theory reduction. This, though of the greatest interest to metaphysics, is again only one aspect of the question. And, even though restricting themselves to this respect, philosophers have often been guilty of oversimplification: they have overlooked the technical difficulties met with in most reduction attempts. A typical example is the claim that solid mechanics is reducible to particle mechanics.

The logicians and metamathematicians have made so far the most reliable contributions to the subject. But they could not be expected to cover the whole field, which has several nonformal regions. It behoves the philosopher to put the various points of view together.

1.2. *The Philosopher's Contribution*

The philosophical writings on reduction can be classed into two disjoint sets: those which mention alleged cases of reduction and comment on them without having made sure that they are genuine and without analysing the reduction process, and those which go to the trouble of analysing some such cases and are therefore able to offer insightful remarks (e.g. Nagel, 1961 and Feigl, 1967). In either case the philosophers interested in reduction seem to take it for granted that science teems with successful reductions: that thermodynamics has been completely reduced to statistical mechanics; that rigid body mechanics has been reduced to particle mechanics; that classical mechanics has been reduced to quantum mechanics; that every relativistic theory has at least one and at most one nonrelativistic limit – and so on. Unfortunately this is also the impression given by most popularisation works, notably by elementary textbooks – the only source of information accessible to most philosophers. Alas, this is not the conclusion one can draw from looking at the original literature. In fact, no rigorous derivation of the second principle of thermodynamics is known: only the thermodynamics of the ideal gas – a very special case – has so far been reduced to molecular dynamics. As to rigid bodies, particle mechanics cannot account for their existence, since the "particles" concerned are quantum-mechanical systems and they are glued by fields, which are extraneous to particle mechanics. Nor does quantum mechanics yield classical mechanics in some limit: it retrieves only some formulas of particle mechanics, none of continuum mechanics, which is the bulk of classical mechanics. Finally, some relativistic theories have no nonrelativistic limits while others have more than one. We shall take up these problems later on. Suffice it to say now that no detailed examination of the many alleged cases of theory reduction is available in the philosophical literature, and that none will be forthcoming as long as the technical literature on the subject be ignored.

However, a few philosophical studies on reduction have been fruitful. The most important and influential of them has been Nagel's (Nagel, 1961). According to Nagel there are two kinds of reduction: *homogeneous* and *inhomogeneous*. In the first case the domains of facts of the two theories concerned are qualitatively homogeneous (e.g. both deal with neural nets), while in the second case they are not (e.g. one deals with mental events and the other with neural nets). Correspondingly, in homogeneous reduction all the concepts of the secondary or reduced theory T_2 are present in the primary or reducing theory T_1. Therefore in this case the reduction amounts to a logical derivation of T_2 from T_1. An example of this is the reduction of particle mechanics to the mechanics of deformable bodies. On the other hand, in inhomogeneous reduction two qualitatively different fields of facts are concerned, so that even if a reduction is effected the secondary theory T_2 is not just subsumed under the primary theory T_1. Far from this, here at least one concept occurring in the reduced theory T_2 is absent from the set of basic concepts of the reducing theory T_1. For example, the thermodynamic concepts of temperature and entropy are not present among the basic concepts of the kinetic theory of gases. Therefore no deduction of thermodynamic statements is possible from the latter theory. In order to effect the reduction, additional postulates must be introduced. These additional assumptions, which are contained neither in T_1 nor in T_2, link all the peculiar terms of T_2 to some terms in T_1, whence they can be called theory-linking or bridge hypotheses. Thus in the kinetic theory of gases the relation between the average kinetic energy of the molecules and the temperature must be postulated, and this additional assumption is not a definition but a new synthetic (factual) hypothesis.

So far so good. But once the secondary theory has thus been enriched and properly organised (i.e. formulated axiomatically), its relation to the primary theory becomes a purely logical one. In other words, the homogeneous-heterogeneous distinction is of a historical or heuristic nature: while it occurs in the theory construction stage, it disappears in the metatheoretical consideration of the finished products. Consequently Nagel's pioneer work on theory reduction should be reconstructed and expanded from the axiomatic perspective. For, even if an axiomatic formulation of a theory may not enrich it essentially, it will always clarify it and, in particular, it will facilitate the clear formulation of problems about the theory.

Let us now move on to other aspects of the question: let us make a critical review of what has been done and give an *aperçu* of the tasks before us.

2. ASYMPTOTIC INTERTHEORY RELATIONS

2.1. *The Intuitive Notion: Its Inadequacy*

The usual situation in science is preaxiomatic. Even when two or more competing theories are compared, they are seldom if ever formulated in an orderly fashion. Hence instead of performing a *systematic* comparison of whole theories, one confronts two or more handfuls of typical concepts and statements. This fragmentary analysis is then employed as a launching pad for general conclusions about the logical relations between the theories.

Moreover, the comparison of theories is often restricted to the *asymptotic* values of certain functions or the asymptotic forms of certain statements, as when Riemannian geometry is said to approach Euclidean geometry as the metric tensor tends to a constant diagonal tensor, or when a special relativistic theory SR comes close to the corresponding nonrelativistic theory NR as the particle velocities v concerned are negligible compared to the velocity c of light in vacuo. The amateur metatheoretician will then treat the theory as a whole and furthermore *as if* it were a function, writing ill-formed formulas such as

$$[1] \qquad \lim_{v \ll c} SR = NR$$

and, in general,

$$[2] \qquad \lim_{p \to a} T_1 = T_2$$

where p is some characteristic parameter. But surely this is just a *metaphor*, for a theory is not a function but a set of statements. Moreover, the reduction (of T_2 to T_1) is *not* always achieved as a parameter approaches some limit value.

2.2. *Nonrelativistic Limits: Sometimes Nonexistent, Sometimes Multiple*

It is usually believed that every relativistic theory has exactly one nonrelativistic limit, so that if the latter is taken, all second order and higher order "effects" are lost but on the other hand the bulk of the facts, the first order "effects", are kept. We shall presently show that, while some relativistic theories have no nonrelativistic "limit," others have more than one, so that the belief under scrutiny is false.

Maxwell's electromagnetic theory for matter-free space is a relativistic theory – moreover it was so *avant la lettre* – and one that has no nonrelativistic limit. In fact, the basic equations of this theory contain no mechanical velocity v, whence there is no point in taking the limit of the functions involved for $v \ll c$. And, as for taking their limit for c approaching infinity, it makes no sense either for it leaves us with the subtheory of static fields, wiping out the peculiarity of electromagnetism – electromagnetic induction. In short, there is no nonrelativistic approximation of Maxwell's electromagnetic theory: there are only the nonrelativistic subtheories of electrostatics and magnetostatics, and the nonrelativistic approximations of electrodynamics (which is a different story). This simple metatheoretical result is important because it explodes the myths (*a*) that relativity is just a matter of higher order "effects" (a quantitative refinement necessary only for high energy phenomena), and (*b*) that every relativistic theory has exactly one nonrelativistic "limit" that covers essentially the same ground

As to the existence of multiple nonrelativistic limits, the simplest case is the one of general relativity, or *GR* for short. *GR* goes over into *SR* for vanishing gravitation (equivalently: for flat space), but it goes over into the classical theory CG of gravitation (of Newton and Poisson) for weak static fields and slow motions. (Actually there is a third limit, namely for a vanishing matter tensor. In this case spacetime can still be Riemannian, and no previous physical theory is obtained, for there are neither matter nor electromagnetic fields left. But this case seems to be without a physical interest, both because it corresponds to no real situation and because it agrees with no previous physical theory: it is a factually empty limit.)

It is of some interest to note that the CG limit of GR is not obtained by letting c go to infinity in all formulas, In fact, one of the specialisations made for obtaining this classical (or rather semiclassical) limit is that all the coefficients of the matter tensor are set equal to zero except the 00 component, which is set equal to $m_0 c^2$. The existence of two different and nonempty limits of GR is also of interest in that it vindicates Einstein's claim (disputed by Fock) that GR is a generalisation of SR; but Fock, too, is thereby shown to be partly right in claiming that GR is a generalisation of CG. As long as the one-limit tenet is kept, either Einstein of Fock will be regarded as possessing the whole truth concerning the nature of GR. Finally, if a quantum theory of gravitation were successful, it would presumably have at least two different limits: GR for either $h \to 0$ or $T_{\mu\nu} = \langle T_{\mu\nu}^{QM} \rangle$, and relativistic QM for vanishing gravitation.

As to Dirac's quantum theory of the electron, there are two ways of obtaining a nonrelativistic "limit" of it. One is the standard procedure of neglecting all the operators whose eigenvalues (or whose averages) are of the second order in v/c or higher; the other is to keep these operators while dropping the "small" components of the state spinor, i.e. those which are of the order of v/c times the "large" spinor components. Not surprisingly, two entirely different "limits" are obtained: the first procedure yields essentially Pauli's nonrelativistic theory of the spinning particle, while the second procedure leads to an equation containing a spin-orbit term absent from the former. This second limit appears to be factually empty. Which refutes one more popular tenet, namely that every "limit" of a given theory covers a subset of facts of the former. As to the first "limit" (Pauli's theory), it reduces to Schrödinger's theory upon dropping the spin operator. The situation so far is summarised in the diagram on the following page. No arrows have been placed after SR and QM because the relations of these general theories to the more special ones are supposed to subsume are so far not well understood. In particular,

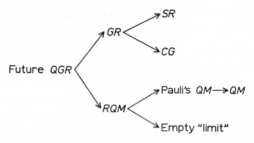

it is not known how to obtain the whole of classical mechanics (i.e. continuum mechanics) from either *SR* or *QM*, even though every textbook, hence almost every philosopher of science, takes these reductions to be a *fait accompli*.

2.3. *The Asymptotic Theory May Not Coincide With The Older Theory*

We have just exploded, by means of counterexamples, the textbook myth that every relativistic theory goes over into a single nonempty classical theory as $c \to \infty$ (or, better, for $v \ll c$). What is more, the resulting nonrelativistic approximation may retain some typically relativistic terms, so that it could not possibly agree in detail with the corresponding classical theory. We saw this for the $GR \to CG$ transition. Special relativity presents a similar case: in the slow motion approximation the total energy of a particle reduces to the rest energy $m_0 c^2$ instead of vanishing, as it should if, in fact, special relativistic dynamics agreed with classical dynamics for small velocities. Moreover, the weaker theory may contain features totally alien to the stronger theory. Thus the symmetry laws (and the corresponding conservation equations) characteristic of *SR* have no counterpart in *GR*, for the Riemann spaces are devoid of overall symmetries. In other words, the weaker theory may not be included in the stronger one even though the two will have a nonempty intersection – for otherwise the very concept of theory strength would be inapplicable.

It would seem, then, that rather than having to do with couples of theories, a classical theory *C* and a revolutionary theory *R*, we are actually confronted with these plus a set *NR* of nonrevolutionary "limits" of *R* – where 'revolutionary' stands for "relativistic," "quantum-mechanical," or perhaps some future kind of theory. The relations between these three theories, regarded as sets of formulas, would seem to be the following:

[3] $NR \subset R$ and $NR - C \neq \emptyset$.

These exceedingly modest metatheorems, however plausible, have not been proved, not even in a single case. And yet such formulas, rather than the ill-formed formulas [1] and [2], do make sense and could conceivably be proved – not however before axiomatising the theories concerned.

2.4. *The Classical Limits of Quantum Theory: Not Well Known*

The situation is even more complicated in quantum theory. In this case one can make the following comparisons: (*a*) quantum theoretical eigenvalues vs. possible classical values; (*b*) quantum theoretical averages vs. possible classical values; (*c*) quantum theoretical operators vs. classical dynamical variables. The first two comparisons are not as easy to make as is usually believed. To begin with, which classical theory is to be taken: classical particle mechanics, classical continuum mechanics, classical electrodynamics, or what? Then, which limits should one take? Should one set Planck's constant equal to zero – and then lose the spin, which has a classical partner? Or should one take very large masses – which makes no sense for a microsystem? Or, finally, should one take the large quantum number approximation – which makes sense only for bound states (discrete spectra)? As to the dynamical variables themselves, all one gets is some analogies which are heuristically fertile and psychologically comforting but not much more than this. The quantum-classical comparison, in short, is far from being a simple affair.

One of the difficulties with the comparison is that the infinitely dimensional Hilbert space, representing the states of the system, has no classical limit. In this respect QM is much more radically new than any other nonclassical theory. (Only the phase of the state vector of a system does look classical, in that its equation of evolution is similar to a classical Hamilton-Jacobi equation. But then the latter need not concern a mechanical system.) If one focuses on the state vector while forgetting the operators, he will tend to interpret QM as a field theory, whereas if he focuses on the dynamical variables he will tend to interpret it as a quaint theory of odd particles. But, clearly, these are just partial classical analogues: the theory as a whole fails to have a classical analogue.

Moreover, QM and CM were not built to cope with the same problems: the former was not framed to pose and answer questions of kinematics, such as the trajectory of an electron in a slit system (the classical "measurement" discussed qualitatively in the first pages of textbooks and then quickly forgotten). The task before the builders of QM was essentially to account for the very existence, structure, and spectra of atoms. The rest – a peculiar dynamics, molecular theory, and nuclear theory – came as a bonus. Consequently the founding fathers of QM did not enlarge mechanics, the science of motion: they conceived a radically new theory. The new theory was called *mechanics* probably due to the mistaken beliefs (*a*) that any hamiltonian theory is mechanical and (*b*) that a fundamental theory must be a sort of mechanics – rather than, say, a field theory. And yet, the foundations of QM are often discussed in the light of (imaginary) experiments concerning the motion of "particles" through slit systems. No wonder such discussions are barren.

Be that as it may, the reduction diagram of the quantum theories of matter, basic quantum mechanics QM and quantum mechanical statistics QMS, are often said to look like this:

$$QMS \rightarrow QM$$
$$\downarrow \qquad \downarrow$$
$$CSM \rightarrow CM$$

where 'CSM' and 'CM' stand for classical statistical mechanics and classical mechanics respectively. (For a number of statements and diagrams like this see Tisza, 1962 and Strauss, 1970.) Unfortunately no one seems to have *proved* that such relations do obtain. To begin with, no rigorous proof of the reduction of CSM to CM is available (see, however, Section 2.6 for an attempt in this direction). Nor is there any proof that QM does go over into CM. The only available proofs concern a few isolated statements, such as Ehrenfest's theorems and some formulas involving total quantum numbers. But this falls short of a systematic proof of the whole theory. Moreover, although QM is usually compared to classical *particle* mechanics (for nowadays only engineers are familiar with the whole of mechanics), it seems obvious that it should rather be compared to *continuum* mechanics, both because of the occurrence of boundary conditions and because, in the relativistic quantum theories, stress tensors can be defined. Also, unlike the quantum field theories and unlike CM, QM presupposes and employs Maxwell's classical electromagnetic theory. Hence it could not possibly go into CM in any of the "classical limits" discussed above, unless the further restriction to null fields were made – in which case the very existence of bodies could not be accounted for. Finally, it is possible to argue that QM is a limit of CM enriched with certain stochastic assumptions concerning e.g. a random force exerted on the system by the environment (de la Peña/Auerbach, 1969). In short, we know very little about the QM-CM relations. And it is a mistake to pretend that we understand them, for this prevents any serious investigation of the matter.

2.5. *The Stochastic-Deterministic Relation*

Other things being equal, a stochastic theory is logically stronger than the corresponding nonstochastic theory or theories: $NS \subset S$. The nonstochastic "limit(s)" of a stochastic theory can be obtained, in principle, in either of the following inequivalent ways. One is to set all the probabilities occurring in the stochastic theory equal to either 0 or 1. A second line is to take the various probability distributions to be concentrated at their averages. A third method is to replace all the random variables by nonrandom ones, for example to substitute

[4] $$\frac{dX}{dt} = kX \quad \text{or} \quad X_{t+1} - X_t = kX_t$$

for the probabilistic formulas

[5] $$\frac{dp}{dt} = kp \quad \text{or} \quad p_{t+1} - p_t = kp_t.$$

There is, of course, no guarantee that either method will yield a reasonable result, i.e. a weaker theory that will work at least to a first approximation. In particular, for the second method to work, the averages must be really stable or nearly so. And yet only the first method will yield a theory contained in the given stochastic theory. Indeed, in this case the weaker theory is obtained without tampering with the basic concepts, while the other two methods involve changing the very nature of some of the basic concepts: they yield not just a specialisation of the given stochastic theory but radically new theories. Whence they are likely to be much more useful than the first method.

The case of the alleged reduction of thermodynamics to statistical mechanics deserves a special section.

2.6. *The Reduction of Thermodynamics: Programme, Not Fact*

The textbook paradigm of theory reduction is, of course, the alleged reduction of thermodynamics to statistical mechanics. This is usually accomplished, or rather attempted, by enriching the basic equations of classical *point* mechanics (wrongly supposed to account for the behaviour of atoms and molecules) with stochastic hypotheses concerning chaotic initial conditions – or, rather, about the irrelevance of the precise initial state. It would be surprising if this trick were to work in general, for one knows that atoms and molecules are not structureless point masses but enormously complex quantum-mechanical systems glued by fields, which are nonmechanical entities.

As a matter of fact the trick does not work in general: in fact, only elementary kinetic theory – which ignores the 2nd law of thermodynamics – and some thermodynamic formulas have been obtained in this way. Thermodynamics as a whole, and particularly the 2nd law, which is its most distinctive feature, has not been reduced to particle mechanics – nor, for that matter, have been fluid dynamics, the mechanics of deformable bodies, and other branches of continuum physics. The reduction of thermodynamics is not a fact but a programme.

Moreover, there is no agreement among specialists as to how a successful reduction of thermodynamics could be accomplished in general – not just for gases in very special pressure and temperature ranges. One possibe line of attack is to try to obtain thermodynamics and other theories of bulk matter from *CM* without the help of any of the usual auxiliary stochastic hypotheses, by showing that the latter are redundant, being entailed by the basic mechanical laws of motion. This is Grad's thesis (Grad, 1967). In particular, Grad claims that it is unnecessary to introduce random perturbations coming from the external world to explain irreversibility – the way Blatt, Kac and others have proposed. The addition of accessory (usually stochastic) hypotheses such as the ones of molecular chaos, and that the prior probability is proportional to the volume in phase space, is regarded by Grad as convenient and possibly inevitable in the present state of the art, but as dispensable in principle, for randomness is born from the interplay of numerous entities of a kind rather than having to be injected from the outside. The difficulties in proving that this is so, i.e. that the laws of motion are sufficient to reproduce all the stochastics feature, would be just technical: they would only concern the handling of large systems of differential equations, some properties of which approximate random behaviour. If Grad is right, then the reduction of (some chapters) of thermodynamics to mechanics is a homogeneous rather than heterogeneous one. (Recall Section 1.2.)

Now, the rationale of Grad's program seems to be two-fold. One is a purely technical one, namely the unsatisfactory way in which most stochastic assumptions are introduced and the sloppy mathematics involved in most of the approximations. The second reason looks philosophical: the reduction achieved so far (which is partial and even so questionable) is of the heterogeneous kind, while if mechanics is regarded as the basic theory the reduction should be homogeneous, i.e. it should be a straightforward deduction.

In any case, Grad has already obtained some remarkable results, and we should wait to see more of them before passing judgement on his approach to the reduction problem. Yet one thing seems hardly disputable: since the elementary constituents of bulk matter do not behave classically but rather quantum mechanically, bulk matter cannot be accounted for in terms of classical particles, hard spheres, and other classical models. What we should look for is a derivation of continuum mechanics and thermodynamics from QM. This is still a programme even though both physicists and philosophers are mostly under the delusion that it has been accomplished.

2.7. *A Disheartening Conclusion*

The upshot of our quick review of the intuitive or asymptotic notion of intertheory relation is disappointing: the asymptotic relation has not been rigorously elucidated, it is far more complex than usually presumed and, what is worse, it is far from having being established in cases that are popularly regarded as closed. The beautiful reduction diagrams one finds in the scientific and metascientific literature are mostly phoney and at any rate unanalysed.

Let us turn to other, better understood kinds of intertheory relations.

3. FORMAL INTERTHEORY RELATIONS

3.1. *The Possible Formal Relations*

Regarded from a purely formal (logicomathematical) point of view, two related theories may stand in the following relations: (i) isomorphism or, more generally, homomorphism; (ii) logical (but not necessarily seman-

tical) equivalence, (iii) inclusion, and (iv) partial overlapping. (If the overlap is empty, the theories are unrelated.) In order to find out which if any
of these situations obtains in a given case, the theories concerned must
be axiomatised, for otherwise one does not know exactly what is being
compared.

Now, the very first thing to do when presenting an axiomatic foundation of a theory is to exhibit its primitive base or set of basic (undefined)
concepts. Barring elementary or first order theories, which are insufficient
in factual science, the primitive base of a factual theory T expressed in
the language of set theory consists of an n-tuple made up of the following
concepts: a set Σ and $n-1$ basic specific and mutually independent (not
interdefinable) predicates P_i. The set Σ, sometimes a cartesian product of
two or more sets, is the reference class of T, i.e. the collection of systems
T is supposed to be concerned about. And the m-ary predicate P_i^m stands
for the ith property of the members of Σ. More precisely, if $\sigma_1, \sigma_2, ..., \sigma_m$
are in Σ, then $P_i^m(\sigma_1, \sigma_2, ..., \sigma_m)$ holds in T or it does not hold in T, and if
it does and if T is factually true, then the formula holds also for the
things themselves. (This characterisation of the base of a factual theory is
naive, as it involves the concept of total truth. But its extension to the
partial truth case, which is the realistic one, need not concern us here.)

Consequently, given two theories, T_1 and T_2, their systematic comparison begins by comparing their primitive bases

[6] $B(T_1) = \langle \Sigma_1, P_1 \rangle$ and $B(T_2) = \langle \Sigma_2, P_2 \rangle$

where the P's now designate whole bunches (actually sequences) of
predicates.

3.2. *Isomorphism and Homomorphism*

Two theories are isomorphic (homomorphic) if there is a one-one (many-
one) correspondence among their respective reference classes and predicate sets such that the structure of these basic concepts is preserved, i.e.
so that sets be made to correspond to sets, unary predicates to unary predicates, and so on. The precise nature of such a correspondence depends on
the structure of the basic predicates, so that no general definition of isomorphism (or of homomorphism), i.e. one that will fit every possible factual theory, can be given. And every special definition requires the prior
axiomatisation of the theory, for otherwise its primitives will not be indi-

vidualised. (The precise form of the axioms is irrelevant for the purposes of proving isomorphism or homomorphism: what is essential is that the primitive base be given and the gross structure of its components be sketched.)

Now, there is but a single case in the physical literature in which the isomorphism of two theories has been claimed. This is the one of wave mechanics (or Schrödinger's "picture" of QM) and matrix mechanics (or Heisenberg's "picture" of QM). However, the available proof is far from rigorous, for any proof of isomorphism requires both the previous axiomatisation of the theories concerned and the introduction of an *ad hoc* definition of theory isomorphism – none of which was available when the isomorphism proof was presented forty years ago. That proof was then heuristic rather than formal. Moreover, there is the suspicion, voiced by Dirac in recent years, that the two theories are not equivalent. Which, if true, should be one more warning that foundations research problems should not be approached in an amateur way.

3.3. *Equivalence*

Two theories with different primitive bases and moreover definitely heteromorphic, may still share all of their formulas. Hamiltonian and Lagrangian dynamics are in this case; although their structure differs on account of their different primitive bases, their formulas can be translated into one another if only the suitable translation code is supplied (e.g. $H = p\dot{q} - L$). In other words, as sets of formulas these theories are the same theory. This holds, of course, for any two different formulations or representations of the same theory: though possibly heteromorphic, they are logically equivalent.

3.4. *Inclusion or Formal Reduction*

T_2 is a *subtheory* of T_1 (equivalently: T_1 is an *extension* of T_2) if (a) T_2 is a theory, i.e. a set of formulas closed under deduction – which not every subset of T_1 will be, and (b) all the formulas of T_2 are also in T_1 but not conversely. To put it another way, let $T_1 + T_2$ be the union of T_1 and T_2 in the sense of Tarski (Tarski 1956). I.e. $T_1 + T_2$ is the set of logical consequences of T_1 union T_2. Then we may say that

[7] T_2 is a subtheory of $T_1 =_{df} T_1 + T_2 = T_1$,

i.e. T_2 adds nothing to T_1. In other words, T_2 is, as a set, included in T_1 just in case T_1 entails T_2 without further ado, i.e. without the adjunction of subsidiary hypotheses. We see then that homogeneous reduction in Nagel's sense (recall Section 1.2) coincides with inclusion.

Neither of the above definitions of theory inclusion is effective as a *criterion* for establishing theory inclusion, for they are concerned with infinite sets of formulas. We are forced then to fall back on the primitive bases of theories, which are finite sets: in fact, they are n-tuples. (Recall Section 3.1.) Roughly, T_2 can be said to be a subtheory of T_1 iff (i) the primitive base of T_2 is contained in the one of T_1, and (ii) every axiom of T_2 is a valid formula of T_1. More precisely, T_2 is called a *subtheory* of T_1 just in case (i)$B(T_2) \subseteq B(T_1)$ (recall formulas [6] and (ii) for every basic predicate P_i^m in T_2, if $P_i^m(\sigma_1, \sigma_2, ..., \sigma_m)$ holds in T_2, so it does in T_1.

(In general the two relational systems $B(T_1)$ and $B(T_2)$ will not be similar in Tarski's sense (Tarski, 1954). Hence a necessary condition for one of them to be a subsystem of the other will not be met even if the subtheory relation in our sense does hold. That is, it is sufficient but not necessary, for T_2 to be a subtheory of T_1, that $B(T_2)$ be a subsystem of $B(T_1)$.)

3.5. *Persistent, Restricted and New Constructs*

There are three possibilities for a construct (concept or statement) in relation to the various extensions of a given theory.

(*a*) *Persistence*: the construct present in the weak theory belongs also to every extension of it (Robinson, 1956). Example: the velocity concept in *CM* and in the non-quantal extensions of it. (As we saw in Section 2.4, *QM* cannot be regarded as an extension of *CM*.)

(*b*) *Extension*: the construct is expanded from one theory to the next: if a function, it is defined on a wider domain or assigned a wider range; if a statement, its intended reference is enlarged. Example: the mass concept in relativistic mechanics as compared to that in classical mechanics.

(*c*) *Emergence*: the construct is newly introduced in one of the extensions of the weak theory, in such a way that it has no partner in the latter. Example: the field concept is emergent with respect to classical mechanics.

It follows that, in order to obtain a *subtheory* of any given theory, either or all of the two following moves can be tried.

(*a*) *Restrict* one or more of the original functions to a narrower domain

– e.g. replace the continuous set representing a body by a collection of isolated points, and accordingly specialise the density functions to deltas.

(b) *Drop* some of the primitive concepts altogether and delete the axioms in which they occur – e.g. drop the stress tensor (rather than setting it equal to zero) as a step towards retrieving particle mechanics from continuum mechanics.

No similar tactics are available for finding the extension of a given theory. What we do have is a collection of heuristic rules, which may or may not work, for relativising nonquantal theories and for quantising nonrelativistic theories. But they do not concern us here. We must now speed forward, to the nonformal intertheory relations.

4. SEMANTIC INTERTHEORY RELATIONS

4.1. *The Presupposition Relation*

Every scientific theory is "based" on some other theories, both formal (logical and mathematical) and nonformal. Thus geometrical optics is based on Euclidean geometry (as well as on other theories), in the sense that the former makes free use of it – actually it contains the whole of Euclidean geometry. To say that a theory A is *based on* another theory B means that A presupposes B, i.e. that B belongs to the background of A. And a theory A *presupposes* another theory B just in case the following conditions are met (Bunge, 1967c):

(i) B is a necessary condition for the meaning or the verisimilitude of A, because A contains concepts that are elucidated in B, or statements that are justified in B, and

(ii) B is not questioned while A is being built, worked out, criticised, tested, or applied – i.e. B is taken for granted, *pro tempore*, as far as A is concerned.

The presupposition relation has then three sides: a logical and a semantical aspect (both taken care of by condition (i) above) and a methodological side. The latter is easiest understood: one never questions everything at a time, but does the questioning step by step. As to the logical and semantical aspects of the presupposition relation, they can best be brought to light upon axiomatising a theory A, for the zeroth step in this process of reorganisation and tidying up is the exhibition of the entire background B of A. If this were done more often, scientific theories would be better understood. Thus it is only when relativistic kinematics is given an axiomatic formulation, that one realises that Maxwell's electromagnetism is prior to it, for without this background special relativistic kinematics is neither meaningful nor true (Bunge, 1967b). If this fact concerning intertheory relations were better known, we would not be flooded with books on relativity that start either with classical mechanics or with the Lorentz transformations rather than Maxwell's equations.

4.2. *Presupposition and Priority*

The above notion of theory presupposition is related to the weaker concept of *theory priority* as sketched by Church (Church, 1962). Thus logic is prior to mathematics in a weak sense, for it supplies a linguistic framework for mathematical discourse and it keeps mathematical inferences under control. But – *pace* logicism – logic is not prior to mathematics in the *strong* sense that it suffices to build mathematics: indeed, every mathematical theory, even the poorest (e.g. the theory of partial order) has at least one extralogical predicate. On the other hand set theory is, so far, prior to nearly all the rest of mathematics in a *strong* sense, for it supplies the basic specific bricks (e.g. the concepts of set, n-tuple, and function) employed in building nearly every mathematical theory. (Prior to the birth of category theory it was possible to hold that the whole of mathematics is reducible to set theory.)

Note that the semantic concept of presupposition does not coincide with the pragmatic or psychological concept of priority. Thus mathematics presupposes logic from a semantic point of view but mathematics usually comes first both historically and methodologically, in the sense that it has motivated most of modern logic and that it still provides the major control and the chief justification for logical research. Quite often, the semantic relation of presupposition runs counter to the pragmatic or historical direction. Thus although particle mechanics came before continuum mechanics, the latter does not presuppose the former but rather the other way around.

Note also that the concept of presupposition is to be kept distinct from the one of entailment, whether syntactic ($\vdash$) or semantic ($\vDash$). If A is deducible from B then obviously A presupposes B in our sense, for B is a supposition under which A holds. But the converse need not hold: A may not follow from its background B alone – and as a matter of fact in general it does not. Thus set theory, which presupposes logic, is not entailed by the latter. Likewise mechanics does not follow from mathematics alone, and relativistic kinematics requires postulates of its own in addition to those of classical electromagnetic theory.

4.3. *Recognition of the Presupposition Relation*

Whether a given theory presupposes another can best be found out by axiomatising at least the former. Otherwise the semantic dependence of one theory upon the other may escape us. Thus it is often maintained that the scattering matrix theory is independent of quantum mechanics and moreover that it should replace the latter. Yet even if the actual computation of the scattering matrix $S_l(k) = \exp[i2\delta_l(k)]$ could always be performed without the help of quantum mechanics (which is not the case), the latter would still be necessary to *interpret* the various mathematical properties of S as physical properties of the system or process concerned. Take, for instance, the most obvious mathematical property of S: its analyticity (as a function of the momentum k) in the upper plane except along the imaginary axis. In order to discover the meaning of the poles of S one examines the asymptotic solution of the Schrödinger equation (the core of quantum mechanics) for the scattering by a finite-range central field, i.e.

$$\underset{r \to \infty}{u} \to (A/r) \sin\left(kr + \delta_l + l\,\frac{\pi}{2}\right) = (B/r)\,[e^{-ikr}\,e^{-il\pi/2} - S_l(k)\,e^{ikr}\,e^{il\pi/2}].$$

For $k = i\varkappa$, with $\varkappa > 0$, $u \to e^{-\varkappa r}/r$, which – according to quantum mechanics – concerns a bound state at the point $i\varkappa$. But since this is the state of a two-component system, we have also this further interpretation: a pole of the scattering amplitude represents a compound system ("particle"), so that the whole S matrix may be regarded as a model (a model object) of a compound system. We owe this discovery to the pre-existing theory of quantum mechanics, which has therefore acted as a *meaning supplier* (Bunge, 1964). If the S-matrix were to become formally self-contained, i.e. self-sufficient rather than dependent on the Schrödinger theory, this semantical relation of presupposition would be regarded as a historical accident, for the theory would stand on its own feet. But since there is so far no satisfactory independent axiomatisation of the scattering matrix theory, it cannot be claimed that the latter is self-sufficient. Moral: First axiomatise, then state your claims about the semantic dependence, or independence, of a theory *vis à vis* another theory.

4.4. *Meaning Changes: Kuhn's and Feyerabend's Theses*

Even if the formulas of one theory reduce to most or even all the formulas of another theory, and even if the two do have the same reference class – i.e. if they are about the same things – they may not have exactly the same *meanings* for, if the two theories are different, they will say different things about their referents. Thus Einsteinian and Newtonian dynamics share most (not all) of their statements for low velocities, but the terms involved in them do not have the same meaning in all cases. And this change in meaning cannot be remedied, because it is rooted to a difference in structure: thus, while distances are frame-dependent in relativity, they are frame-independent in classical mechanics.

Hence Kuhn (Kuhn, 1962) is quite right in pointing out that Newton's laws of dynamics are *not* derivable from Einstein's: that it is not just a matter of quantitative agreement in the nonrelativistic limit, but of a "displacement of the conceptual network." Only, Kuhn put forth his thesis in a misleading way, namely by asserting that "the physical referents" of the Einsteinian laws differ from the Newtonian ones, so that in the attempt to retrieve the latter from the former "we have had to alter the fundamental structural elements of which the universe to which they apply is composed" (p. 110). This would mean that the two theories are not about the same thing – which is plainly false, as the two are about particles. Kuhn's thesis is right if reformulated in the following way. In a scientific revolution both the form and the content of some concepts change. Sometimes a conceptual change corresponds to a change in referent (e.g. the replacement of continuum theories by atomistic theories of matter), at other times the referent is kept (though not the theoretical model of it) but there is a meaning change. (Which, incidentally, reinforces the thesis that the reference of a construct is only one of the two components of its meaning – the other being its intension.)

Feyerabend's well-known thesis on meaning changes (Feyerabend 1962) is more radical and less defensible than Kuhn's. "What happens when transition is made from a restricted theory T_2 to a wider theory T_1 (which is capable of covering all the phenomena which have been covered by T_2) is something much more radical than incorporation of the *unchanged* theory T_2 into the wider context of T_1. What happens is rather a *complete replacement* of the ontology of T_2 by the ontology of T_1, and a corresponding change in the meanings of all descriptive terms of T_2 (provided these terms are still employed)." This thesis has a grain of truth but, as it stands, it is half-baked and even inconsistent. It is half-baked because it contains two key concepts that are not elucidated by its author: one is the concept of theory coverage (which can be explicated [Bunge, 1967c]), the other is the concept of meaning (and the associated concept of ontology of a theory) – which also can be elucidated (see the next subsection). It is a pity that such a revolutionary thesis should have been stated with the characteristic sloppiness of traditional philosophy.

Worse: taken literally, Feyerabend's thesis is *self-contradictory*, for a theory cannot be pronounced wider than another and at the same time incommensurable with it in point of meaning. Indeed, if the change in semantics ("ontology") were as complete as Feyerabend claims, then surely the two theories would not be comparable as to scope: they would merely talk about different things. Consequently we would be unable to ascertain which of them has the larger coverage. Nevertheless, as I said before, there is a grain of truth in Feyerabend's thesis: namely, that scientific progress carries with it changes in meaning. Yet even such changes, though occasionally radical, are not always as radical as Feyerabend thinks. Feyerabend's own favorite example bears out this contention.

Indeed, when Feyerabend claims that "It is [...] impossible to define the exact classical concepts in relativistic terms" (Feyerabend, 1962, p. 80), it is obvious that he disregards the elementary concept of restriction of a function, which often does the trick. For example, the relativistic concept M_R of mass, which can be introduced via certain postulates, enables one to define the classical concept M_C of mass, namely thus:

[8] $M_C = {}_{df} M_R \mid B \times U_M$,   where $M_R : B \times K \times U_M \rightarrow R^+$.

Here, '$M_R \mid B \times U_M$' stands for the restriction of the map M_R to the set $B \times U_M$ while the domain of the revolutionary concept M_R is the set of ordered triples $\langle b, k, u \rangle$ with b in the set B of bodies, k in the set K of physical reference frames and u a member of the set U_M of mass units. Likewise with other concepts that get relativised to the reference frame and so become joint properties of physical systems and frames.

In conclusion, scientific revolutions are not as wild as "cultural revolutions", and the thesis of the meaning changes associated with scientific revolutions is important enough to deserve careful philosophical elucidation. (For further criticisms see Coffa, 1967 and Nagel, 1970.) To this task we now turn.

4.5. *Elucidation of the Concept of Meaning Change*

In order to clarify the concept of meaning change associated with theory replacements, we must start by elucidating the very concept of meaning. A possible explication of the latter is offered by the following definition, which encapsulates what I call the synthetic view on meaning, for it combines intensionalism with extensionalism.

Let c be a concept, proposition, or theory. We define the *meaning* of c as its sense or connotation together with its reference or denotation (Bunge, 1972c). In short,

$$\mathcal{M}(c) =_{df} \langle \mathcal{S}(c), \mathcal{R}(c) \rangle,$$

where the sense $\mathcal{S}(c)$ equals the set of formulas entailing c or entailed by c, whereas the reference $\mathcal{R}(c)$ is the collection of objects to which c refers – quite apart from whether it refers to them correctly. The set of constructs entailing c may be called its *purport*, and the collection of constructs entailed by c its *import*. The larger the import of a construct, i.e. the less its dependence upon other constructs, the greater its importance. Primitive constructs have the largest import.

Once we have the concept of meaning it makes sense to elucidate the concept of meaning change associated with the replacement of a construct c by another construct c'. We define the *meaning change* accompanying such a replacement as the ordered pair

$$\delta_{\mathcal{M}}(c, c') =_{df} \langle \delta_{\mathcal{S}}(c, c'), \delta_{\mathcal{R}}(c, c') \rangle,$$

where the first coordinate is the *change in sense*:

$$\delta_{\mathcal{S}}(c, c') =_{df} \mathcal{S}(c) \, \varDelta \mathcal{S}(c'),$$

and the second coordinate the *change in reference*:

$$\delta_{\mathcal{R}}(c, c') =_{df} \mathcal{R}(c) \, \varDelta \mathcal{R}(c'),$$

where the triangle designates the symmetric difference operation defined in set theory.

Let now T be a theory replaced by another theory T'. There will be a net meaning change only in the following cases: T is a subtheory of T' in the sense of Section 3.4, T and T' overlap partially, or they are totally disjoint. But the latter case is as uninteresting as the other extreme, namely the case when T and T' are just equivalent formulations (Section 3.3) of one and the same theory. Since for any pair T, T' of theories we shall have to do with infinite sets of statements, the change in meaning may seem to be quite unmanageable. This difficulty can be dodged by restricting the whole affair to the axiom bases of T and T'. Consequently the above formulas will be regarded as concerning the sets of postulates of T and T' respectively. But, of course, this will not be welcome by the lovers of fuzziness, to whom axiomatics is a real menace.

5. PRAGMATIC INTERTHEORY RELATIONS

5.1. *Heuristic Relations*

Pragmatic relations may appear among scientific theories in more than one way, sometimes because they are sought but most often unexpectedly. The main kinds of pragmatic intertheory relations seem to be the following: (*a*) *heuristic*: one theory suggests or helps to build another theory; (*b*) *methodological of the first kind*: one theory is instrumental in devising empirical tests of another theory; (*c*) *methodological of the second kind*: one theory (an "established" one) is regarded as a condition another theory (a new one) must satisfy, usually in some "limit."

The ways one theory can suggest the construction of another are numerous and they are reluctant to strict classification, for they depend not only on the theories themselves but also on the frame of mind of the theoretician. One man will search for inspiration in mathematics, another will try generalising in a purely formal way, while a third theoretician will reinterpret a given scientific theory and a fourth one will pursue certain analogies that others had failed to "see." However, a couple of general points can be made.

A first point is that a heuristic relation is often, in a sense, the converse of a logical relation. Thus although particle mechanics is a subtheory of continuum mechanics, the actual process (or rather attempt) of constructing theories of fluids and solids has often gone from particles to particle systems to continuous bodies. In general, in attempting to construct a richer theory one will usually step on the available theories, which one may wish to convert into subtheories of the new one.

A second point is that the heuristic scaffolding using ideas borrowed from pre-existing theories should be critically examined and discarded if necessary once the new theory has been built. Otherwise it may become an obstacle to a correct statement, hence understanding, of the new theory. Suffice it to recall that the Faraday-Maxwell theory was not adequately understood until the beginning of this century, partly because it had been dragging mechanical analogies.

A third point is that a powerful theory may be a source of inspiration not only for more advanced theories but also for the overhauling of previous theories. Thus mechanics was seen in a new light after field theory was constituted, to the point that continuum mechanics can be treated as a field theory (Truesdell and Toupin, 1960). It is well known that solid state physics has been able to benefit from the mathematical machinery of quantum electrodynamics. It is less well known that classical electrodynamics can be made to account for some quantum effects if the zero point energy, a typical quantum term, is included (Marshall, 1963, 1965 and Boyer, 1969a, 1969b). And even second quantisation can be mimicked within classical theory (Schiller, 1967; Bourret, 1967). Of course all of this is hindsight, whence it cannot be used to prove that classical physics is enough and that it alone can explain the supposed mysteries of quantum physics. What it does show is that new theories do not just pile up on top of old theories: in the process of growth the whole network of physics gets transformed.

5.2. *Empirical Tests of One Theory with the Help of Another*

No matter how close to experience a certain theory may seem to be, its empirical test will require the help of several other theories entering the design of the test as well as the design and reading of the scientific instruments involved in the test. In other words, in any experimental situation two sets of theories (or scraps of such) will become involved (see Chapter 10):

(1) the theory to be tested (the *substantive* theory), and

(2) a collection of fragments of theories accounting for the experimental set-up (the *auxiliary* theories).

The two sets of theories may have disjoint reference classes: thus a theory concerning the condensation of cosmic dust will have to be tested with the help of telescopes and other instruments designed with the help of some scraps of optics and mechanics. As new experimental techniques are introduced, unexpected pragmatic relations of this kind come into being. Surely Newton was unaware of the electronic and computer equipments currently being employed in testing certain applications of his theory of motion and gravitation (e.g. lunar theories).

That no theory suffices to design and interpret its own tests, seems obvious from the many-sided character of measurements, yet it is tacitly denied by all those who regard quantum mechanics either as concerning only experimental situations (e.g. Bohr) or as providing all the necessary materials to build a general quantum theory of measurement that would in turn provide an exhaustive account of every possible experimental situation (e.g. von Neumann). If either thesis were true, quantum mechanics would be the sole theory in need of no auxiliary theories for its tests. But experimenters seem to think otherwise: they regard quantum mechanics as in principle susceptible to falsification by experiments, and experiments as framed in the light of a bunch of more or less clearly stated ideas borrowed from a number of theories. In short, the quantum theories are no exception to the rule that the empirical test of any scientific theory calls for the intervention of several other theories, so that no scientific theory is methodologically isolated from the rest of science. Which is just as well, for otherwise there would be no mutual control.

5.3. *Empirical Tests of One Theory Through Another*

Some theories are not empirically testable in a direct way, not even when conjoined with auxiliary theories (in the sense of Section 5.2), but must be tested *via* some other theory. For example, there is at present no known way of testing relativistic thermodynamics, which renders it operationally meaningless. Never mind, for the theory is cherished for the sake of completeness. Yet there should be a means for checking some formulas of the theory. For example, we should know whether temperature transforms like a length (the usual view) or like an energy (the correct view if the relation to statistical mechanics is recalled). Since no measurements are currently available to decide this point, one must look elsewhere for a vicarious empirical test. Relativistic statistical mechanics provides it to the extent that it entails relativistic thermodynamics – which it does only fragmentarily. (Recall Section 2.6.) But this theory is not directly testable either, although one hopes to get very soon extremely high temperature and jet velocity data relevant to it. The way to test relativistic statistical mechanics is to subject relativistic mechanics to empirical tests. This is an incomplete test, for the auxiliary stochastic assumptions are not separately tested. Moreover, it involves several auxiliary theories. But this is how things stand: the empiricist ideal of the theory that faces alone empirical data, because it has an empirical content, is just a philosophical myth.

5.4. *Theoretical Tests*

Every new promising theory is subjected not only to empirical tests but to purely conceptual tests as well. The conceptual test of a factual theory consists, essentially, of an examination of the way the theory manages to cope with the valid tradition – both scientific and philosophic. Even a revolutionary theory, if scientific, will not rebel against everything but will be consistent with logic, most if not all of mathematics, and a number of factual theories regarded as true to a first approximation. (The rumour started by von Neumann and propagated by a few mathematicians and philosophers, according to which quantum mechanics involves a revolution in logic, is groundless: quantum mechanics, when axiomatised, proves to presuppose certain mathematical theories that have ordinary logic built into them. Besides, if quantum mechanics did obey a logic of its own, it could not be conjoined with classical theories, e.g. Maxwell's, to derive testable statements.)

If the new theory covers entirely new ground, one not previously treated by a previously accepted theory, then it should only be required to be *compatible* with the bulk of the background knowledge. But if the reference class of the new theory includes the reference class of a less comprehensive theory, and if the latter had been found particularly true, then a stronger condition will be placed on the new arrival. The latter will be required to *include* the old theory (in the sense of Section 3.4), or at least to have a *sizable overlap* (note the deliberate vagueness) with it in some "limit" or other. Ideally, the new theory should have all the virtues but none of the vices and limitations of the older one.

The condition that the new, more comprehensive theory should give back the sound parts of the theory it intends to supersede, is often called the *correspondence principle*, and is usually credited to Bohr. Bohr was perhaps the first to state it explicitly in relation to quantum theories and the first to exploit it systematically, although the rule had been employed earlier, notably in checking (conceptually) special and general relativity. It is intended to be a general principle subsuming all those principles employed in a preliminary theoretical test. But, as shown in Section 2, not every theory complies with it.

Bohr and his followers have regarded the special correspondence principle employed in building and checking quantum mechanics as a quantum theoretical law. This betrays a superficial analysis of scientific laws, all of which are supposed to concern objective patterns rather than pairs of theories. In other words, correspondence principles are *metatheoretical and heuristic*, not intratheoretical principles (Bunge, 1961). If they were primary laws, rather than metalaws, they would allow us to make predictions. In any case, the intervention of such metanomological statements in the evaluation of scientific theories shows once more that theories are assayed in the light of both facts and ideas. A number of criteria, some of them of a philosophical kind, have to be satisfied by any new theory in addition to factual adequacy. (See Bunge, 1967c, Chapter 15.)

6. QUEER VIEWS ON INTERTHEORY RELATIONS

6.1. *The Popular View*

The popular view on intertheory relations is, like every other popular view, simple enough: it holds that every historical sequence of scientific theories is *increasing*, in the sense that every new theory includes (as regards its extension) its predecessors. On this view nothing is ever lost: whatever is added remains as a permanent gain, and moreover the process converges to a limit that is the union of all the successive theories. This view can be made to look plausible by choosing extremely short subsequences that happen to conform with it. These are, of course, the subsequences occurring in the standard textbooks, which record only successes, never failures, and state without proof that the more successful theories contain (actually or asymptotically) their less fortunate predecessors.

The popular thesis is philosophically superficial, as it neglects the semantical aspects (the meaning changes referred to in Sections 4.4 and 4.5), and it is false as a historical hypothesis concerning the advancement of science. Moreover, it mixes up logic and history, two poles that ought to be kept apart – but so do two other views on intertheory relations, the Copenhagen and the dialectical views, to which we now turn.

6.2. *The Copenhagen View*

According to the Copenhagen view, quantum mechanics is not a more comprehensive theory than classical mechanics (by which only particle mechanics is meant). The ground offered for this contention is that it would be meaningless to speak of a microsystem, say an atom, as a thing in itself: according to Bohr and his followers (Bohr, 1958a; Feyerabend, 1968), one should always speak of an entire block mysteriously constituted by a microsystem, the measurement set-up (even when we are dealing with atoms in outer space?), and the subject in charge of the experimental arrangement. The reason for this seems clear: we have no (experimental) access to the microsystem except through an apparatus manipulated by someone. Now, the apparatus is to be described in classical terms: it is a macrosystem. Therefore, the argument concludes, quantum mechanics presupposes classical mechanics and even the whole of classical physics. As a standard textbook (Landau and Lifshitz, 1958) puts it at the very beginning: "quantum mechanics occupies a very unusual place among physical theories: it contains classical mechanics as a limiting case [not true: recall Section 2.4], yet at the same time it requires this limiting case for its own formulation." We saw earlier that quantum mechanics does not contain the whole of classical mechanics but only a tiny fragment of it. Let us now examine the second thesis.

This muddled view has two roots: classicism and positivism. Rather than admitting that the referents of quantum mechanics are (or rather were) unheard-of-entities, so much so that they do not satisfy the law statements of classical physics, the classicist will try to go on using classical analogies – such as those of position, momentum, particle, and wave. Never mind if this leads him to contradictions such as those of talking about the diffraction of particles and the collision of waves: he will en- shrine absurdity as a principle – the principle of complementarity. The second root of the Copenhagen view that quantum mechanics presupposes classical mechanics is even more obviously wrong: it is the Vienna Circle confusion between *reference* and *test* – a confusion cleared up quite some time ago (Feigl, 1967; Bunge, 1967b). Surely in order to test quantum mechanics, or any other physical theory, one needs some fragments of classical physics: recall the role of auxiliary theories in testing substantive theories (Section 5.2). But this does not entail that, when formulating quantum mechanics, one has to start with classical mechanics – only to end up by concluding that the two are really mutually inconsistent. Nor does it entail that it would be meaningless to speak of a microsystem apart from a measurement device. Quantum electrodynamics speaks most of the time of free electrons, and when computing energy levels of atoms and molecules one never takes any apparatus into account: the apparatus coordinates simply do not occur in most of the formulas of the quantum theories. In short, although quantum mechanics is tested with the help of theories that are not quite consistent with it, these do not occur in its formulation. The Copenhagen view on intertheory relation is, in sum, one more confusion that must be cleared away.

6.3. *The Dialectical View*

Dialectical philosophers have maintained that the historical succession of ideas has been a dialectical process whereby every new idea has as-similated its predecessors and overcome the latter's inner contradictions, while at the same time containing its peculiar contradiction – the prime mover that would eventually lead to its own dialectical negation. Every successful new theory would hold to its historical antecedents the relation of dialectical sublation or *Aufhebung*, in the sense that it would somehow contain its predecessors though not in a "mechanical" way (not as sub-theories).

It is true that the awareness of incompatibilities, and in particular contradictions, is a major source of scientific progress – not however because scientists love contradiction but rather because they cherish con-sistency, both internal and external (i.e. consistency of the given theory with the bulk of human knowledge). But this does not establish the dialectical thesis. Firstly it is by no means proved that every scientific theory must contain some contradiction. True, transition theories – such as the elastic theory of light and Bohr's classical quantum theory – do sometimes contain contradictions, but nobody is happy with them when they are discovered. Secondly the opinion that every successful new theory overcomes and in a way subsumes some of the old theories is overly optimistic. Sometimes the new theory is definitely shallower than the one it competes with but is accepted because it has some other advantages – witness the case of thermodynamics versus the atomistic theories of the second half of the last century. Moreover, we cannot exclude the possibility that, in abeyance to an obscurantist philosophy, new but inferior theories may come to replace some of the present ones: theoretical progress, however needed to improve our understanding and mastering of reality, is by no means a logical or a historical necessity.

But history aside, the philosophical trouble with the dialectical view on intertheory relations is that it is fuzzy, because the *Aufhebung* relation has not been analysed. Moreover, it apparently resists analysis in terms of logic or mathematics, as dialectics is nonformal and its nucleus is ontic contradiction (tension, strife). Nor is the converse explication, of logic in terms of dialectics, possible. For, although dialecticians have often claimed that formal logic is a sort of slow motion approximation of dialectical logic, the latter has never been formulated explicitly and has never been shown to entail formal logic. Besides, the whole idea of a dialectical *logic* adequate to account for a dynamical world rests on a Presocratic confusion between logic and ontology: at best, dialectics can claim to be an ontological and/or epistemological theory. In any case, the *Aufhebung* relation has not been clarified and therefore the dialectical view on intertheory relations is itself obscure: it is something to be explained rather than an explanatory theory. This is why it has made no contribution to the study of logical, semantical and methodological relations among scientific theories – least of all when combined with positivism and the Copenhagen doctrine (as in Strauss, 1970): the composition of obscurities does not yield clarity.

7. CLOSING REMARKS

No general theory of intertheory relations seems to have been proposed so far. We have got only (*a*) a calculus of deductive systems, model theory and category theory, which combined take care of the formal relations among theories, and (*b*) a set of scattered remarks on the nonformal relations among theories. These remarks are mostly sketchy and informal, and very often incorrect. Not only do we lack a systematic treatment of intertheory relations – aside from the formal side of the question – but the detailed analyses of specific theory pairs are scanty and marred by a number of textbook myths. Worse: we are caught in a circle: there is no general theory because we do not have enough detailed studies of particular cases, and such studies are scarce because there is no general theory that can be applied to them.

And yet it is clear that we do possess some of the major tools for attempting to perform a systematic analysis of the relations among theories, chiefly the above mentioned calculus of deductive systems and model theory, and axiomatics. Amateur analyses, which neglect using these tools, can at most produce some valuable hints. For it is only well ordered systems, with a definite structure and a comparatively perspicuous content, that can be compared with profit. Moreover, since theories are infinite sets of statements, only their axiomatic foundations and a few typical theorems are manageable. Hence axiomatisation is a prerequisite for an accurate analysis of the logical and semantical relations among theories. This applies, in particular, to the problem of the reducibility of one theory to another. As Woodger said several years ago – without however catching the attention of the philosophers of reduction – "Strictly speaking we can only fruitfully discuss such relations between theories when both have been axiomatised, but outside mathematics, this condition is never satisfied. Hence the futility of much of the discussion about whether theory T_1 is reducible to theory T_2 'in principle'. Such questions cannot be settled by discussions of that kind but only by actually carrying out the reduction, and this is not done and cannot be done until the theories have been axiomatised" (Woodger, 1952).

As long as the preposterous tenet is held, that a scientific theory is not a hypothetico-deductive system but an inductive synthesis, a metaphor, or what not, and as long as an irrationalist reluctance to axiomatics is felt, no decisive advances can be expected in the study of intertheory relations. And as long as neither careful case histories nor a general theory are available, we should refrain from pressing intertheory relations for philosophical juice.

NOTE
* Some paragraphs are reproduced from Bunge (1970c) with the permission of the editor and publisher.

CHAPTER 10

THE THEORY/EXPERIMENT INTERFACE*

A scientific theory can make contact with experience in at least three ways: (*a*) it can be tested for factual truth by means of experience (observation, measurement or experiment); (*b*) it can be used to plan and interpret observations, measurements or experiments; (*c*) it can be employed to practical (noncognitive) ends such as making an artifact or destroying it. We shall deal here with the first two kinds of contact and shall approach the problem from a general methodological point of view without getting involved in the technicalities of statistical inference and experiment design: our purpose will be primarily philosophical, namely to stress the intimate interlocking of theory and experience, that refutes the claim that either of the two poles is overriding.

We shall show that the test of a scientific theory is a complex process in which the following stages can be recognised:

(i) the theory is subjected to preliminary tests of a nonempirical nature, such as compatibility with an accepted body of knowledge;

(ii) the theory is adjoined subsidiary assumptions till specific predictions are derived, and objectifiers or indices of the unobservables concerned are hypothesised;

(iii) fresh data are produced (rather than gathered) with the help of a set of auxiliary theories;

(iv) these data are confronted with the theoretical predictions and both are evaluated.

To work.

1. FIRST COME THE NONEMPIRICAL TESTS

1.1. *Agreement with Fact not Decisive*

According to the official philosophy of science, agreement with fact is not only necessary but also sufficient for the acceptance of a scientific theory, because scientific theories are just data summaries or at worst codifications of data and slight extrapolations from them. According to this view, if a theoretical prediction conflicts with an empirical datum it is the former, not the latter which has to go – and, indeed, without any appeal, for experience is the highest court of appeal. This view is methodologically, philosophically and historically untenable. First, because it is standard scientific practice to reject data that conflict with established theories. Second, because data are anything but given: they are produced and interpreted with the help of theories. Third, because most theories do not concern observations and measurements, let alone acts of perception, but things or rather idealised models of them. Fourth, because – as we shall see – testable propositions seldom if ever follow from the assumptions of a single theory but, rather, are usually entailed by the theory in conjunction with additional assumptions and with bits of information other than those serving to check the theory – just as the generalisation "All men are mortal" is insufficient to conclude that Socrates is mortal.

The received view is also refuted by the history of science – a kind of experience philosophers of science should always keep in mind. Indeed, the history of science abounds in examples of theories that have been upheld in the face of adverse empirical evidence – and rightly so, for the data proved wrong in the end. This was the case with the "anomalies" in all of the planetary motions except Mercury's: they were not interpreted as refuting Newton's celestial mechanics but as pointing to the incompleteness of the available empirical information or to the difficulty of making exact calculations with this theory. It was also the case for certain delicate measurements, performed by competent experimentalists, that seemed to refute the constancy of the velocity of light and thereby both classical electrodynamics and special relativity. And it is the case with every new theory accounting for a sizable subset of the set of available data even though it conflicts with some of them, provided no better theory is in sight: the discordant evidence is then declared to be an insignificant residue, or at worst a sad fact of life – when not simply false. Such was the case with Einstein's theory of Brownian movement, which was decisive in establishing the atomic theory of matter. Indeed, the theory had been confirmed by the measurements of J. Perrin but it had been refuted by the equally delicate (but, as it turned out, ill-interpreted) measurements of V. Henri (Brush, 1968). It was accepted, among other reasons, because it explained Brownian movement (even though it was doubtful that it predicted it accurately) and because it squared with other theories, such as the kinetic theory of gases and the chemical atomic theory. In any event, agreement (disagreement) with fact is seldom sufficient to accept (reject) a scientific theory.

1.2. *Four Batteries of Tests*

Whether we like it or not, every organic body of scientific ideas is evaluated in the light of the outcomes of four batteries of tests: metatheoretical, intertheoretical, philosophical, and empirical. The first three constitute the nonempirical tests and all four together can give us a hint as to the viability or degree of truth of a theory (Bunge, 1961b, 1963, 1967c).

A *metatheoretical* examination is one bearing on the form and content of a theory: it will seek, in particular, to establish whether the theory is internally consistent (no petty task), whether it has a fairly unambiguous factual meaning as formulated, and whether it is empirically testable with the help of further constructs, especially hypotheses relating unobservables (e.g., causes) to observables (e.g., symptoms). An *intertheoretical* examination will try to find out whether the given theory is compatible with other, previously accepted theories – in particular those logically presupposed by the theory concerned. This compatibility is often attained in some correspondence limit, e.g., for large (or small) values of some characteristic parameter such as the mass or the relative speed. A *philosophical* test is an examination of the metaphysical and epistemological respectability of the key concepts and assumptions of the theory, in the light of some philosophy. Thus if positivism is adopted, phenomenological theories – such as thermodynamics, *S*-matrix theory and behaviouristic learning theory – will be favored while theories concerning the composition and structure of the system concerned will be neglected or even fought without regard to empirical evidence and to the thirst for deeper explanation. I am not advocating philosophical censorship but recalling that, as a matter of historical record, this kind of consideration is always made – sometimes for better, often for worse (Hertz, 1956; Margenau, 1950; Bunge, 1961b, 1967c; Kuhn, 1962; Agassi, 1964).

If a theory is believed to comply with the accepted metatheoretical, intertheoretical and philosophical requirements, it may get ready for some empirical tests. (Whether it does in fact conform to the canons is another matter. And whether it will succeed in provoking the curiosity of a competent experimentalist, is yet another matter.) An *empirical* test is, of course, a confrontation of some of the infinitely many logical consequences of the initial assumptions of the theory, enriched with subsidiary hypotheses and with data, with some information obtained with the help of observations, measurements or experiments designed and read with the help of the given theory and of further theories. Thus in order to test a gravitational theory one will focus on some of its theorems and will build, with some of the concepts of the theory, a model of the physical system concerned, that will incorporate only the relevant features of the real thing; the next step will be to design and execute certain measurements bearing on that model and based on theories such as optics and mechanics.

1.3. *Priority of Nonempirical Tests*

No theory is given an empirical examination unless it is believed to have passed all three batteries of nonempirical tests. Most often some of these tests are not actually performed, either because they are exceedingly difficult (as is the case with consistency tests) or because it is intuitively felt that the theory satisfies the nonempirical requirements – an impression that very often proves to be wrong. The incompleteness of such tests does not diminish their value and it does not refute our contention that the nonempirical tests precede the empirical ones. In any event, demonstrably inconsistent theories can be written off with hardly any qualms, and completely off the track theories are seldom, if ever, considered for empirical tests. No matter how original it is, a scientific theory must be "reasonable" and "likely": it must be well built, it must not go against the grain of justified scientific beliefs, and it must not postulate items that are either metaphysically objectionable (such as an electron's ability to make decisions) or epistemologically opaque (such as a hidden variable with no possible overt manifestation).

In all three nonempirical tests consistency is involved: internal consistency, the consistency with other pieces of scientific knowledge, and the consistency with philosophical principles. Consistency is not only a logical virtue but also a methodological one. Indeed, an internally inconsistent theory may predict anything and may therefore be confirmed by mutually conflicting pieces of evidence. And a theory that fails to cohere with other theories will not be able to enjoy their support and suffer their control – as is the case with many pseudoscientific ideas. The worst that can happen to a theory is not that it be refuted by experiments it has induced itself, but that it remain hanging in mid-air with neither friends nor foes.

As to the consistency of our scientific theories with the dominant philosophy and even the whole of our world view, we care for it because philosophy is indeed relevant to scientific research and in particular to the selection of research problems, to the formation of hypotheses, and to the evaluation of ideas and procedures. It goes without saying that subservience to a wrong philosophy can be harmful to research; thus intuitionist philosophy has blocked the advance of psychology in some countries, chiefly Germany and France. But it is a fact that consistency with the dominant philosophy is always sought or appreciated – and even believed to obtain when in fact it does not, as was the case with the relativistic and atomic theories in relation to positivism. This makes a critical examination of philosophical principles the more necessary. But the adjustment between science and philosophy should be mutual rather than one-sided – under penalty of stiffening both partners. The fact that a happy and fruitful marriage of philosophy and science is needed makes it even more desirable. At any rate, although there are nonscientific philosophies, scientific research is permeated with a number of philosophical ideas (Bunge, 1967c).

2. SECOND STAGE: THEORY GETS READY FOR CONFRONTATION

2.1. *Theories Untestable in Isolation*

One century ago the great Maxwell remarked that, when setting out to test candidates for law statements, one does not rush to the laboratory but starts by doing some further theoretical work: "the verification of the laws is effected by a theoretical investigation of the conditions under which certain quantities can be most accurately measured, followed by an experimental realisation of these conditions, and actual measurement of the quantities" (Maxwell, 1871). Note the three stages: experimental design (a piece of theoretical work), construction of the set-up, and performance of the empirical operations. (For details see Bunge, 1967c.) The experimental design will involve further hypotheses concerning the links of a given magnitude (e.g. gas pressure) with one that can be measured (e.g. the length of a liquid column), as well as a theoretical representation of the whole set-up. The same applies, *a fortiori*, to the process of verification of systems of hypotheses, i.e. theories.

It is impossible to subject a scientific theory to empirical tests without roping in further theories. For one thing, while every theory covers some aspects of its referents (e.g. its magnetic properties), any empirical operation involves real objects that refuse to abstract from all those aspects which every theory deliberately neglects. Secondly, a theory may be untestable by itself for failing to concern observable facts: it may restrict itself to making assertions about what happens or can happen, whether or not the events are observable. (But it may still have a factual content even though it may have no empirical content.) Thus a theory of electric circuits is about electric currents but it does not state the conditions of its own test: the latter requires a further theory, namely electrodynamics, which will bridge unobservables such as the current intensity to observables such as the deflection angle of a meter. In most cases we do not need a full theory but just a few scraps of various theories.

To put it another way: scientific theories are *untestable by themselves* both because they are partial and because they involve transobservational concepts that are not linked, within the theories, to any empirical concepts. These links, indispensable to test a theory, must be borrowed from some other area of knowledge. Thus a psychological theory will become testable to the extent to which objectifiers (behavioural, physiological, neurological, etc.) can be adjoined to it. In sum, if we wish to see how our theories fare empirically, we must call in additional ideas instead of eliminating every theoretical element by way of "operational definitions".

2.2. *Adding a Theoretical Model of the Referent*

The adjunction of fragments of other theories is necessary but insufficient to obtain results comparable with data: since in experience we handle individual things – a given liquid body rather than the body genus, this human subject rather than mankind, and so forth – we must add *subsidiary assumptions* concerning the relevant details of the system concerned. Thus in the case of a theorem in electromagnetic theory we must add special hypotheses and data concerning the shape, charge distribution and magnetisation of the field sources.

A general theory does not contain such subsidiary assumptions precisely because it is general. It is a comprehensive framework compatible with a whole family of sets of subsidiary assumptions. Every such set sketches a *theoretical model* of the thing concerned. Any such model is cast in the language of the theory although it is not dictated by the latter. Clearly, a theoretical model may, but need not, be visualisable: being constructed with the concepts of a theory, it will be as abstract (epistemologically speaking) as the theory itself. Thus classical mechanics is consistent with a large variety of models of planetary systems; likewise, it is consistent with many models of liquids: the continuous medium model, the gas-like model, the crystal-like model (Ising's), and so on. A general theory cannot be tested apart from some model or other – as long as the model is regarded as a theoretical image of the thing concerned rather than as a heuristic metaphor.

2.3. *The Importance of Specific Hypotheses*

A subsidiary hypothesis concerning some trait of the object of study may mask the truth value of a general theory, particularly if few data are available, as is often the case in a new area of research. For example, suppose there are two rival theories concerning the Q-ness of bulk matter – an imaginary physical property. Each theory hypothesises its own functional relation between this peculiar property Q and the area A of the thing concerned. The first theory assumes that (in appropriate units) $Q = \frac{1}{2}A^{1/2}$ while the second postulates that $Q = (2/A)^{1/2}$. Suppose further that measurement yields the following bits of information: (*a*) e = The linear dimensions D of the experimental object are of the order of unity; (*b*)e^* = The value of Q as measured on the experimental object is 1.0 ± 0.2. Unfortunately the shape of the thing is not observable: it must be guessed. This is where a subsidiary assumption must be adjoined: in order to set the theory in motion we must hypothesise a model of the thing – in this case a visualisable model of an unseen thing. Suppose the following situation occurs:

$$e : D = 1$$

$H_1 : Q = \frac{1}{2}A^{1/2}$	$H_2 : Q = (2/A)^{1/2}$
S_1: The thing is a disk	S_2: The thing is a sphere.
$H_1, S_1, e \vdash Q_1 = \pi^{1/2}/4 \cong 0.4$	$H_2, S_2, e \vdash Q_2 = (2/\pi)^{1/2} \cong 0.8$

Clearly the right-hand result is consistent – within experimental error – with the measured value of Q, i.e. 1 ± 0.2. But it would be folly to write off H_1 on this ground – for, by replacing S_1 by S_2, we would come up with $Q = \pi^{1/2}/2 \cong 0.9$, which is an even better value of Q than Q_2. This case is imaginary of course but by no means artificial. Moral: Watch the model object, for a good model can save (temporarily) a poor general theory, just as an inadequate model can ruin (permanently) a good general theory.

2.4. *Assuming Models and Seeking Them*

Theoretical scientists can be found to state, in prefaces and in concluding remarks, that every scientific theory is "based on" experimental data. But on reading the work sandwiched between empiricist covers, one finds that it does not fit this philosophy. One finds in fact that – unless it consists in a new theory – the work either (*a*) computes quantities that may (sometimes) be subsequently confronted with empirical results, or (*b*) combines given experimental data with a general framework in order to infer some specific feature of the system concerned. In either case work starts from some general framework rather than from scratch, if only because that general framework will suggest the kind of information to be sought in the laboratory or in the field. Thus energies and scattering cross sections, rather than precise positions – or, for that matter, entropies and stresses – will be computed or measured in the case of the scattering of atomic beams, because the general theory says that the former quantities are relevant.

More precisely, in theoretical science there are direct problems and inverse problems. A *direct problem* looks like this: Given both a general framework and a specific theoretical model of the system concerned, find either a general formula of a certain kind or an instance thereof. Here go some examples from physics. (*a*) Given classical mechanics (general framework) and a definite fluid model (determined, say, by a certain distribution of masses, stresses and forces), compute the trajectory of an arbitrary particle in the fluid (i.e., a streamline). (*b*) Given quantum mechanics (general framework) and the standard model of the helium atom (a three-body system kept together by Coulomb forces), deduce the energy spectrum. (*c*) Given the same general theory as in (*b*) and the usual model of a target as a central field of force, compute the scattering cross section for a beam of given characteristics.

The corresponding *inverse problems* would be these. (*a*) Given classical mechanics and a set of streamlines, infer the mass and force densities as well as the stress tensor. (*b*) Given quantum mechanics and a sample of an energy spectrum, guess the constituents of the system and the forces among them. (*c*) Given quantum mechanics and a cross section *vs.* energy curve, infer the interparticle forces. In every case the inverse problem is: Given a general theoretical body and certain empirical data, find the model which best fits both.

To put it symbolically, the general theory supplies a function f that relates the hypothesised model m to a testable consequence t, i.e. $t=f(m)$. Thus in the case of the direct scattering problem, t may be a phase shift and m the assumed hamiltonian (equivalently, the interaction forces). An inverse problem, on the other hand, boils down to finding the inverse f^{-1} of f, so as to obtain: $m=f^{-1}(t)$. The effective inversion of f calls for the determination of the suitable information t, as well as applying or inventing a suitable mathematical technique. In no case is empirical information alone given, let alone sought: the very kind of information the experimentalist runs after is more or less suggested by the general framework. As a well-known specialist in scattering problems remarks, "The most easily accessible [experimental scattering] information helps us not at all if we are not astute enough to find a procedure for obtaining the hamiltonian from it" (Newton, 1966).

If the general theory is consistent and the direct problem is properly formulated, and at all solvable, it will have a unique solution. Not so with most inverse problems, which are characteristically indeterminate (Bunge, 1963). This holds particularly for the problem of finding a model on the basis of a general framework and a set of data: usually the two determine jointly a whole class of models (e.g., hamiltonians) rather than a single model. To realise the indeterminateness peculiar to inverse problems (e.g. finding a model) we need not go into the ambiguities encountered in elementary particle physics (see e.g. Newton, 1966). We find it in elementary problems such as the one of determining the intensity and the voltage of an alternating current from measurements, which give only average values.

2.5. General Schema

Call T_1 the theory to be tested and S_1 the set of subsidiary assumptions added in order to derive some statements T'_1, specific enough to come close to experience. S_1 will include a theoretical model of the system(s) under consideration and it may include simplifying assumptions such as linearisations. The theory T_1 – an infinite set of statements – will be judged on the performance of the theorems T'_1, which are not only finite in number but also partly alien to T_1 even though they are cast in the language of T_1. (One more reason for refusing to identify 'theory' and 'language'.) Notice that the real situation, in which T_1 and S_1 jointly entail T'_1, is a far cry from the standard view according to which T_1 single handed yields T'_1, which should in turn be directly comparable to the empirical evidence.

As a rule not even the T'_1 will be directly testable, for they will involve theoretical concepts such as the one of stress (whether mechanical or psychological) that have no empirical counterpart. In order to connect T'_1 with experience we must adjoin a further batch of hypotheses, namely the objectifiers or indices of the unobservable entities and properties in question. Thus gravity is objectified by motion and appetite by amount of food consumed. Call I_1 the set of indices or objectifiers employed in bridging the gap between the theory T_1 and experience. These indices are not "operational definitions" but full blown hypotheses that should be checked independently even though they may go unquestioned in the process of testing T_1. They are hypotheses devised on the basis of the available knowledge A as well as of T_1 itself – for the theory under considertion must decide which kind of evidence will be relevant to it. At any rate, once the inventive process is over it must be possible to show that the objectifier hypotheses are well founded: that A and T_1 jointly entail I_1.

We still need some particular empirical statements if we are to derive specific predictions. Call E_1 the set of data fed into the theory. In order to introduce them into T_1 we must translate them into the language of T_1. For example, astronomical data, originally expressed in geocentric coordinates, will have to be translated into heliocentric coordinates. This data translation is done with the help of T_1 itself and of some fragments of the antecedent knowledge A. Let us call E_1^* the set of data couched in the language of T_1 and ready to be fed to it. In a careful logical reconstruction, A, T_1, I_1 and E_1 will entail E_1^*.

Finally, from the particular theorems T'_1 and the translated data E_1^* we will obtain a set T^* of testable consequences – not just of the theory T_1 under examination but of T_1 in conjunction with all the remaining assumptions and data. T^* will face the fresh empirical evidence produced in order to test T_1.

In brief, the preparation of the theory T_1 for empirical testing is as follows.

Construction of a model of the referent	S_1
Deduction of particular theorems	$T_1, S_1 \vdash T'_1$
Construction of indices	$A, T_1 \vdash I_1$
Translation of data	$A, T_1, E_1, I_1 \vdash E_1^*$
Drawing testable consequences	$T'_1, E_1^* \vdash T^*$

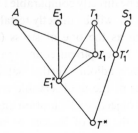

Fig. 1.   The derivation of testable consequences of T_1 involves the antecedent know-ledge A, some data E_1, a model S_1, and bridge hypotheses I_1.

3. THIRD STAGE: NEW EXPERIENCE IS PRODUCED AND PROCESSED

3.1. *Interpreting What We See*

The next task is to produce a set E^* of data relevant to the theoretical predictions T^*. The performance of this task often calls for a theoretical work comparable in volume with that conducted in the previous stage.

Consider the X-ray diffraction pictures, the main empirical tool of analysis for molecular biologists. These pictures make no sense whatever except in a theoretical context: what one actually sees are dark spots and rings around a center. Such patterns bear no obvious relation to the spatial configuration of the atoms in the crystal; theory alone tells us the meaning of these (natural) signs. What one does in order to "read" such pictures is to hypothesise a given atomic configuration (call it T_1) with the help of several fragments of physical and chemical theories. Further, one admits that electrodynamic theory (call it T_2) accounts for the nature and behaviour of X-rays. From T_1 and T_2 one computes (with the help of Fourier analysis) the theoretical diffraction pattern, i.e. the one that should obtain if both T_1 and T_2 were true. But this pattern is invisible: we need, in addition, some bridge to the observed picture. Diffraction patterns may be rendered visible by means of sensitive photographic plates. The mechanism of this process is explained by a third theory, namely photochemistry, which will be called T_3. An X-ray diffraction picture (a blind *datum*) becomes *evidence* for or against a molecular structure theory T_1 when it can be deduced from it with the help of auxiliary theories (electromagnetic optics and photochemistry), one of which explains the diffraction mechanism while the other explains the blackening mechanism. In short, T_1, T_2 and T_3 jointly entail E (see Figure 2).

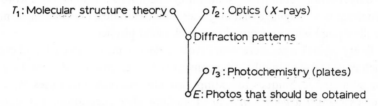

Fig. 2.   An X-ray diffraction picture makes sense if it can be be predicted on the basis of a model of the crystal and with the help of two auxiliary theories: one accounting for the nature of X-rays, the other for the blackening process.

The experimenter will, of course, start at the other end: he will produce E and will proceed to try to guess T_1 with the help of the theories T_2 and T_3, which he will take for granted in this particular context. His is an inverse problem (see Section 2.4). When the crystal is very complex, as is the case with a protein, which contains thousands of atoms, his guesswork is very intricate – so much so that only a small fraction of X-ray diffraction pictures have so far been deciphered. But he can always get some help from resemblances with cases studied before. Moreover, to make a start he may deliberately discard much empirical information: he may begin with a low resolution instrument, just as the astronomer often starts with a low power telescope. Unless he makes such simplifications he may get no pattern at all – and a pattern is of course what he is after. Just as a rough theoretical model is better than no model at all, so, digestible data are preferable to data indigestion.

The crystallographer's task would be greatly simplified if theoretical chemistry were more advanced: if it were possible to deduce all the possible configurations that any given set of atoms could fit. Such a detailed calculation of possible molecular configurations requires a fourth theory – quantum chemistry – which has been around for four decades but is still not quite prepared to undertake such a formidable task. If and when such a breakthrough is accomplished, the logical tree in Figure 2 will have to be supplemented with a branch descending from quantum chemistry to T_1. The unraveling of the "meaning" of many a presently mysterious X-ray picture depends on further theoretical development rather than on finer observation and measurement techniques.

3.2. *Knowing What We Measure*

The instructions concerning laboratory operations are sometimes worded in a pragmatic language that disguises their theoretical foundation, as may be illustrated with an example from classical physics.

Every precision measurement involves electric measurements, and every such measurement involves the comparison of electric resistances. One of the standard techniques for comparing electric resistances uses Wheatstone's bridge, a *pons asinorum* of modern instrumentation. The design and operation of a Wheatstone bridge are based on the elementary theory of electric networks, the central laws of which are Kirchhoff's and Ohm's. Figure 3 represents in a fairly direct way a theoretical model of the Wheatstone bridge in the equilibrium state, i.e. when no electricity flows through the galvanometer G.

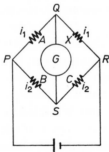

Fig. 3. The Wheatstone bridge, in combination with elementary network theory, allows us to infer X from A, B and C.

Under these conditions, Kirchhoff's second law yields

$$V_{PQ} - V_{PS} = 0$$

for the left fork and

$$V_{QR} - V_{SR} = 0$$

for the right one. In turn, every one of these potential differences is, by Ohm's law,

$$V_{PQ} = Ai_1, \qquad V_{PS} = Bi_2$$
$$V_{QR} = Xi_1, \qquad V_{SR} = Ci_2$$

whence the final formula

$$X = AC/B.$$

(The galvanometer G bridging the points Q and S of the circuit does not occur explicitly as a referent in these formulas, because it registers zero current.)

The preceding formulas may be summarised in the following *physical statement*

P: In one of the branches of the Wheatstone bridge there exists a point S at which the electric potential has the same value as the potential at a given point Q in the other branch.

The technician will employ the following *operational statement* that translates the preceding proposition into the language of human action:

O: If one of the terminals of the galvanometer in a Wheatstone bridge *is connected* to a point Q *chosen arbitrarily* on one of the branches of the bridge, and if the other terminal *is displaced* along the other branch, a point S will be *found* for which the pointer of the galvanometer will be *seen* to come to rest at the zero of the scale.

(The italicised words are of course the pragmatic terms in the sentence.) Although the technician may be satisfied with this operational statement O, the only justification for O is the preceding physical (and theoretical) statement P. Moreover, it is P which led Sir Charles to invent his bridge. (The mere observation that no current passes through G could otherwise be interpreted as indicating that the meter is out of order.) In general, however theory-free ordinary experience may be, no precision experience is possible in science without some theory, even though the description of the experience may not exhibit this dependence upon theory. The more precise a measurement the more complex the underlying theory (Levi, 1947). An analysis of two typical measurements in modern physics will bear out this claim.

3.3. *Measuring Probabilities in Atomic Physics*

In the simplest case – the one studied by philosophers – probabilities are measured by counting relative frequencies. But indirect probability measurements, i.e. measurements via theoretical formulas, are about as frequent. A good example is the measurement of the intensity of a spectrum line as an index or objectifier of a transition probability. (For the concept of index or bridge hypothesis, see Section 2.5.) The link between the two is roughly this: The more probable a transition between two energy levels, the more intense the corresponding spectrum line. If the transition is highly probable, a bright line is seen; if the transition probability is low, a dim one and if the probability is nil, no line at all. (If, notwithstanding the theory, a line is seen where it should be absent, then the corresponding transition is called forbidden – and the suitable correction is made in the theory.)

Since many spectrum lines are visible with the naked eye, the claim could be made that when looking at any of them what is actually being observed is a transition probability. This could do, on condition that it be realised that such an observation is heavily loaded with theory, to the point that without it what would be seen would be just a bright colored stripe. After all, the transitions in question are quantum jumps from one atomic energy level to another, and the probabilities are calculated with the help of theoretical formulas. Moreover, the experimentalist must design the equipment (light source, diffraction grating, photographic plates, wavelength comparator, etc.) in accordance with several theories (notably optics). The latter requires not only the effective realisation of the conditions assumed by the theories involved (e.g., the equal spacing of the grating lines) but also certain assumptions that cannot be controlled exhaustively. Among the latter assumptions the following occur: the arc temperature does not change from one photograph to the next, the arc atoms under study enter the arc stream at a constant rate, and they do not appreciably absorb the light emitted by their kin. Once the empirical data have been collected and sifted (criticised and processed), theory comes in to compute the transition probabilities in terms of the measured quantities. The formula employed to infer such probabilities from the measurement outcomes, is the Einstein-Boltzmann equation. The measurable magnitudes occurring in this formula are the temperature and the light intensity. While the former can be measured with high accuracy, the standard deviation of the measured intensity values is, even today, no less than about 30%. The whole procedure is so complicated and involves so many uncertainties that the first comprehensive and reliable table of "experimental" atomic transition probabilities was published only in 1961 after 30 years of teamwork (Meggers *et al.*, 1961).

3.4. *Measuring Probabilities in Nuclear Physics*

In nuclear physics the probability of an event, such as a nuclear reaction, is usually given by the total cross section for that event: again, because the corresponding theory (quantum mechanics) says so. (See Figure 4.)

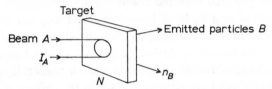

Fig. 4.   Nuclear reaction $A \rightarrow B$. The number n_B of particles emitted is related to the incident flux I_A by the theoretical formula: $n_B = I_A \sigma_{AB} N$, where σ_{AB} is the total cross section for that reaction and N the number of target particles presented to the beam.

In the total cross section the scattering angle is obliterated. Since the scattered intensity depends upon the angle, it is also necessary to consider the differential cross section, or cross section per unit angle and unit energy interval – a higher level concept serving to define the total cross section. In the laboratory one measures the differential cross section referred to the laboratory frame of reference. If this value is to be compared with a theoretical prediction, it must be converted into a center-of-mass value. Thus translated, a measurement outcome may look like this (an actual figure accepted at the time of writing): "At an angle of 20° 8 and an energy of 156 MeV, the proton-proton cross section in the center of mass frame equals 3.66±0.11." (Different teams of physicists will obtain values differing by as much as 15%.) In general, for the scattering of A particles by B particles at the energy E and the angle ϑ_{cm}, one will have a statement of the form

$$\sigma(A\text{–}B, E, \vartheta_{cm}) = n \pm \varepsilon,$$

where n is a (fractionary) number and ε the total error. Note how far from sense experience such a laboratory outcome is: A and B name particle species the members of which are imperceptible: they are objectified by means of instruments embodying several theories. The energy E is measured indirectly and the scattering angle ϑ_{cm} is calculated from the measured angle. Finally the error ε is arrived at with the help of statistics. In sum, the whole experimental procedure is permeated by theoretical ideas, and the very idea of a scattering cross section (as distinct from a geometrical cross section) makes no sense outside microphysics.

3.5. *Empirical Evidence neither Purely Empirical nor Conclusive*

Contrary to popular superstition, science has little use for pure (uninterpreted, theory-free) data, and no evidence is definitive one way or another. Even the data gathered with the naked eye are meaningless unless they can be integrated in a body of knowledge, and they are all subject to uncertainty. One of the archaeologists taking part in the excavations (1967) of what may have been King Arthur's legendary Camelot declared at one point that *he thought he could see up to six or seven* different layers of ruins – stones that would not have been looked for in the absence of the legend. During the 19th century all astronomers *saw* that the nebulae (our present galaxies) were continuous (gaseous) bodies rather than the clusters of stars which the late 18th century astronomers *saw*. And they failed to see what everyone can now *see* by himself, namely the black dust clouds (e.g. in the rings of the spiral galaxies). We do not report what we see with mindless eyes but report, rather, what we *think* we see: scientific observation, unlike the observation of babies and empiricist philosophy, is permeated by hypotheses and expectations, some explicit, most tacit. Even ordinary observation is determined jointly by sensation and ideation; experimental psychology has established that the same sensory input can give rise to different perceptions while in different circumstances different sensations may correspond to the same perception (Hebb, 1966).

Measurement does not eliminate observational uncertainty although an analysis of measurement in the light of mathematical statistics can render the uncertainty precise. This, indeed, is an aim of the calculation of the standard deviation of the random errors of observation. But not all errors are of this kind: apart from the systematic errors incurred in the design or operation of a laboratory equipment, one has to reckon with possible errors in the theoretical part of any indirect measurement. Thus, before the 1920's the size of the galaxies had been found to be roughly ten times smaller than their actual size. Similarly, in the early 1950's all intergalactic distances had suddenly to be multiplied by two, when a mistake was found in the previous calculations. Contrarily, sometimes one knows there is something wrong with the data and cannot ascertain what it is. Thus as recently as 1967 the measured values of the rotation period of Venus went from 5 days with the optical method to 244 days with the radar method (Smith, 1967).

In sum, there are no hard and fast data: there are only hard skulls sheltering the belief in the ultimate character of data. Every experimental technique is based on assumptions that ought to be subjected to independent checks, and the practical implementation of any such technique is subject to conceptual mistakes and perceptual errors, as well as to objective random variations in both the object and the instrument involved. Empirical data are no more certain than the theories relevant to them; but both data and theories, though uncertain, are corrigible.

3.6. *General Schema*

Any empirical operation presupposes a body A of antecedent knowledge. A includes, in particular, a set E_2 of data and a heap T_2 of scraps of theories. Although E_2 and T_2 are criticisable on other occasions, they go unquestioned in the given empirical investigation: they will be taken as authoritative, however far from authoritarianism we may be. On the strength of A, and particularly of T_2, bridge hypotheses I_2 will be devised which will enable the experimenter to objectify unobservables and, conversely, to interpret his readings in theoretical terms. In short, A and T_2 entail I_2.

The next step is to design an observation or experiment, involving I_2, the outcome of which can be relevant to the theory T_1 under test. (There certainly is much poorly designed experimentation but precisely for this reason it is of little worth and, even if aimless, it cannot be totally cut off from all theory.) The experimental design will involve a number of specific subsidiary hypotheses S_2 sketching a theoretical model of the equipment. From S_2 and T_2 certain consequences T'_2 concerning the functioning of the equipment during the empirical operations will follow. In short, T_2 and S_2 jointly entail T'_2.

Finally the operations proper will be performed. Call E_2 their outcome, or rather the empirical reports once cleansed and condensed with the help of the theory of errors. To make sense, the E_2 must be read in terms of both the theory T_1 under test and the auxiliary theory T_2. That is, from T_1, T_2 (or rather T'_2), I_2 and E_2 we shall derive a set E^* of data relevant to T_1.

In sum, we have the following tree:

Constructing a theoretical model of the equipment S_2

Deducing particular theorems $T_2, S_2 \vdash T'_2$

Constructing indices $A, T_2 \vdash I_2$

Translating data $E_2, I_2, T_1, T'_2 \vdash E^*.$

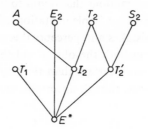

Fig. 5.   The raw data E_2 are cooked and dressed in theoretical terms with the help of antecedent knowledge A, the theory T_2 and the model S_2 of the experimental equipment, bridge hypotheses I_2, and even T_1 itself.

4. FOURTH STAGE: THEORY MEETS EXPERIENCE

4.1. *Statements: Theoretical and Empirical*

We are now in the possession of two sets of comparable statements: the theoretical predictions T^* and the empirical evidence E^*. Our present task is to confront them in order to draw some plausible "conclusion" concerning the worth of the substantive theory T_1 partly responsible for T^*. But before setting out to do this we must realise that T^* and E^*, though comparable, cannot expect to coincide, for they are of different kinds. This point must be stressed in consideration of the standard view according to which the T^* are just consequences of T_1 alone, while E^* might also be contained in T_1, in any case the ideal being the equality of these two sets. (In fairness, the current theories of inductive logic [Carnap, 1950; von Wright, 1957; Lakatos, 1968] do not concern a scientific theory but an isolated hypothesis h and a pure empirical evidence e, and they attempt to compute the degree of confirmation of h and the probability of the conditional "$e \Rightarrow h$" given both the empirical evidence e and the conditional "$h \Rightarrow e$." No actual examples – scientific examples that is – of either conditional are ever mentioned, and the empirical evidence is regarded as sacrosanct. Moreover, no method for assigning probabilities to statements is given.)

Let us insist that T^*, far from being a sample of T_1, is derived from T_1 jointly with a definite model (S_1), some data (E_1) and some bridge hypotheses (I_1). Likewise E^*, far from being a set of bare empirical statements, is a sample of interpreted outcomes of scientific experiences: otherwise it would not be comparable with T^*. Even so, E^* and T^* are not on the same level for, far from concerning the object in itself, any member of E^* refers to a couple object-empirical arrangement. (The Copenhagen interpretation of quantum mechanics pretends that this holds for every theoretical statement of this theory as well, but this is false: (a) the theory may concern free systems, i.e. things that are coupled to no measurement device, and (b) no general theory can account for the idiosyncrasies of every conceivable apparatus.) Change the apparatus or, better, the whole experimental arrangement, and a new set E^* of data is likely to result. If it does not, something may have gone wrong. At any rate, T^* and E^* are not quite on the same footing. The following analysis will make their differences apparent.

A quantitative prediction is a theoretical statement about the value of some "quantity" (magnitude) Q of some real system in a certain state. Actually the system described by the theory is not the real thing σ of the kind Σ the theory intends to account for, but an idealised sketch or theoretical model m of it. (Actually, m is a theoretical model of the referent in a given state.) In a typical case, Q will be a real valued function, so that a prediction of a value of Q will take the form

[1] $Q_s(m) = r$

where the subscript s indicates the scale that has been adopted, while the value r of the function is some real number. (Still better, Q is a real valued function on the cartesian product of the set M of models by the set S of scales.) Thus far the theoretical prediction.

The experimenter handles the real thing σ with a certain experimental technique t which he implements with a certain sequence a of acts. (In microphysics an ensemble of similar systems, rather than a single system, will usually be available. But this is not always the case: thus individual nuclear reactions are "observable.") His results will depend not only on the thing σ but also on his technique t and its implementation a. More precisely, a single measurement result concerning the magnitude Q will take the form

[2] $Q'_s(\sigma, t, a) = r'_a$

where r'_a is again a number (rarely identical with the theoretical value r). (Better still; Q' is a real valued function on the set of ordered quadruples $\Sigma \times S \times T \times A$.) The point to note is that the measured Q' and the theoretical Q are altogether *different functions*. No wonder they seldom have the same values. (Recall Chapter 4, Section 2.2.)

The individual measured values [2] are then processed with the help of mathematical statistics. The two most important outcomes are the standard deviation (a measure of the total error) and the average value, which is taken to be an estimate of the true value. A statement about the average Q' is of the form

[3] $\underset{a \in A}{AV} \, Q'_s(\sigma, t, a) = r'$

where the subscript '$a \in A$' means that the average is taken over a sample A of measurements. (Ideally A is infinite. Actually it is not, whence it is not stable. But its fluctuations decrease with the increasing size of the sample.) In general r' will differ from any of the individual values [2].

Once the average and the error have been computed, the experimenter may wish to go over his raw data [2] once again, to weed out or else justify any anomalous data that may have crept in. These black sheep will be those values lying beyond the limits agreed upon in advance (usually the 3 standard deviations limit). But if too many black sheep are found it will be necessary to proceed to a critical examination of the experimental procedure itself. The experimenter may then find that some of the assumptions have not been satisfied – for example that, contrary to hypothesis, every measurement act has influenced the subsequent act, i.e. that the condition of statistical independence has not been fulfilled. At any rate the experimenter does not accept his own results uncritically: he examines them in the light of both methodological theory (mathematical statistics) and substantive theory (e.g. mechanics). And the theoretician should not claim (as the Copenhagen school people do) that his own predictions concern measured values, for in general he does not know what experimental technique nor which steps to implement it will be adopted.

4.2. *The Confrontation*

Having emphasised that T^* and E^* are separated by a chasm, let us now bridge it. Let E^* be relevant to T^* – for otherwise we may come up against one of the paradoxes of confirmation. Under this assumption there are only two possibilities: either E^* agrees with T^* or it does not. 'Agreement' here means less than identity and more than compatibility. A qualitative prediction such as "The scattered beam will be polarised" may be regarded as confirmed if, in fact, the beam proves polarised if only partially. But if the prediction is quantitative, as is the case with "The degree of polarisation of the scattered beam will be p [a definite number between 0 and 1]", then we need a different truth condition. The one tacitly adopted in physics seems to be the following. Let

[4] $p : P(m) = x$

be a theoretical prediction concerning a model m of the thing σ in a certain state, and let

[5] $e : P'(\sigma, t) = y \pm \varepsilon$

be the outcome of a run of measurements of P, on the real thing σ, with the technique t. The theoretical value is x, the average of the measured values is y, and the statistical scatter of these values is ε. The theoretical prediction p and the empirical datum e may be said to be *empirically equivalent* just in case the theoretical value x and the experimental (average) value y differ (in absolute value) by less than the experimental error ε – a tolerance agreed on beforehand. In short (Bunge, 1967c)

[6] $Eq(p, e) =_{df} |x - y| \ll \varepsilon$

The precise meaning of the inequality relation will depend on the state of the experimental techniques. A theoretical statement and an empirical statement will be said to *agree* with one another if and only if they are empirically equivalent. Clearly, identity is a particular case of agreement.

If the "overwhelming majority" of the data E^* agree with the theoretical predictions T^*, then we declare T_1 to be *confirmed* by that particular set of data. Note, first, that we do not require every datum to agree with the corresponding prediction, and this because outlying data are bound to occur which can usually be discarded. But of course we must keep our mind open to the possibility that some black sheep may actually be white. Note also that the theory under test is declared to be confirmed by a certain set of data, and not just confirmed: this is a reminder that empirical tests, however prolix, are never exhaustive. Thirdly, note that we have not specified how strongly E^* confirms T^*. In actual science no degrees of confirmation are computed: the usual concept of confirmation is a comparative not a quantitative one.

What if, on the other hand, E^* disagrees with T^*, i.e. if there is a sizable subset $E'^* \subseteq E^*$ of data that fail to match the theoretical predictions T^*? According to both inductivists and refutationists we should then reject T^* and also T_1: disagreement with experience refutes a theory and therefore forces us to give it up. However, this is not in keeping with actual scientific practice. In actual science one does not accept unfavourable evidence without further ado but subjects it to a critical scrutiny, for any datum can be distorted by a number of factors. It often happens that the unfavourable evidence E'^* is rejected either because it is inconsistent with veteran theories or because it stems from poor experimental design.

If E'^* is discarded, then there are two possibilities for the theory T_1 under test. If T_1 is a veteran theory, then we shall continue to use it while keeping in mind the anomalous E'^* – for, after all, they might prove not to be calumnies at all. If on the other hand T_1 has not yet proved its worth, while the unfavorable evidence is uncertain, then we should suspend judgement on the truth value of T_1 and wait for a new crop of more reliable evidence.

The negative outcome E'^* should be accepted if the auxiliary theory T_2 has been independently confirmed, if the experimental design passes a critical examination, and if the data are not mostly isolated values that can be discarded by the rules of thumb of mathematical statistics. But the acceptance of the unfavourable evidence E'^*, while committing us to reject the predictions T^*, does not entail the refutation of the substantive theory T_1. Indeed, a number of premises have been used, in addition to T_1, to derive the predictions T^*: the subsidiary hypotheses S_1 (including those sketching a model of the object concerned), the bridge hypotheses I_1, and the data E_1. We are faced with what has been called *Duhem's problem*: Given a set of premises entailing a set of consequences refuted (substantially if not totally) by experience, find the subset of premises responsible for the failure, with a view to replacing them by more adequate ones. This problem seems to be much more important than the problem of devising and computing degrees of confirmation.

In Duhem's view (Duhem, 1914), when a theory disagrees with data, two equally legitimate procedures can be applied. One is to try to save the central hypotheses of the theory by eventually adding some auxiliary assumptions regarding either the referent of the theory or the experimental arrangement. The second way out is to correct some or all of the basic hypotheses, without having the slightest suspicion as to what to correct in the first place nor in what sense. Clearly, rationalists and conventionalists will recommend the first move while empiricists will recommend the second one. But in either case the prospects seem rather bleak.

Our previous analysis of the way T^* is derived (Section 2) confirms the complexity of Duhem's problem but at the same time it suggests that a solution may be possible in every case provided care is taken to list the relevant premises. For, if the adverse empirical evidence is reliable, there are again two possibilities: either T_1 has stood up to tests in the past, or it is a newcomer. In the former case we should keep T_1 temporarily and subject the remaining premises responsible for T^* to a searching criticism. Of all these premises, usually the data E_1 and the bridge hypotheses I_1, though fallible, have been checked previously and in any case they are not usually questioned at the time T_1 is being questioned. Hence the most likely culprits are to be found among the subsidiary assumptions S_1, whether it be the theoretical model or the simplifying assumptions. We should then start by relaxing the latter and/or modifying (usually in the sense of further complication) the theoretical model. It is only after unsuccessfully trying many and widely different models that we must cast serious doubts on the theory T_1. Thus in the case of the current classical theories of liquids, what theoreticians do is to keep trying more and more complex models of the liquid structure while retaining the laws of motion and, in general, the whole framework of classical mechanics.

On the other hand, if the theory T_1 under test is new or nearly so, then we shall subject both T_1 and S_1 to an exacting examination. Yet the suspect premises are not on the same footing: the more specific ones are the most likely to be false, for they take more chances and they are less likely to have been tested. One should therefore start by questioning the subsidiary premises S_1 – in particular the theoretical model – and the more specific axioms of T_1. The more generic postulates of T_1, those which T_1 shares with several other theories, are the least likely to be in need of reform, at least with regard to the domain in which they have been confirmed in the past. (When such extremely general and deep assumptions are shown to be wanting, then whole bunches of theories are likely to suffer reform.) In any case, this search for error need not be haphazard: it should proceed from the newer and narrower to the older and wider. An axiomatisation of the theory under scrutiny should be extremely helpful in this search, for then all the presuppositions and the assumptions of the theory will be on display for all to see. Such an axiomatic organisation of the theoretical material will be particularly useful if the three kinds of premises (the presuppositions, the generic postulates, and the specific ones) are clearly separated. (See Chapters 7 and 8.) One will then begin by replacing the various specific assumptions one by one, and watching the effect of every such change on the testable consequences T^*.

Eventually one will come up with a new body T'^* of theoretical predictions, one which will agree with the total empirical evidence E^* or at least with a sizable part of it. Anything can come out of this readjustment work: a new theoretical model, and/or a slightly different theory, or else a radically new theory – or even a wholly different approach to theory construction. Criticising theories in a constructive spirit, i.e. trying to build better ones, is one of the most rewarding experiences – one that dogmatists and carpers alike are spared.

In summary, when E^* is relevant to T^*, the confrontation process looks like this:

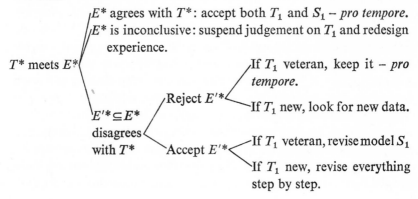

E^* agrees with T^*: accept both T_1 and S_1 – *pro tempore.*
E^* is inconclusive: suspend judgement on T_1 and redesign experience.

T^* meets E^*

$E'^* \subseteq E^*$ disagrees with T^*

Reject E'^*
If T_1 veteran, keep it – *pro tempore.*
If T_1 new, look for new data.

Accept E'^*
If T_1 veteran, revise model S_1
If T_1 new, revise everything step by step.

5. CONSEQUENCES

If the preceding analysis is substantially correct, we must abandon the widespread belief that every theory *single-handedly* faces its empirical jury. Firstly because, in order to describe specific observable facts, a theory must be adjoined some information, a definite model, and a bunch of hypotheses linking unobservables to observables. Secondly because the empirical jury is itself backed up by a body of theory, a further model (of the empirical set-up), and some bridge hypotheses. In short, the theory under test calls for additional hypotheses and past experience, just as the new data intended to check it require some antecedent theory and further special hypotheses: theory lives not on fact alone, and data are not self-sufficient either. This renders them both comparable and mutually controllable.

Consequently it is false that, as the inductivists claim, any theory should in principle *entail* the very same data from which it was induced. Not only are scientific theories not concocted out of pure data, but by themselves they entail none. Therefore theories cannot have any empirical content. Only single hypotheses, such as Snell's law of refraction and Galilei's law of falling bodies, might be said to yield, by mere specification, any number of data – provided at least one item of empirical information is adjoined to them and provided the deep difference between theoretical and empirical statements is overlooked. But the theories to which these two hypotheses belong (wave optics and classical gravitation theory) are not testable just by instantiation. In other words, the conditional "$h \,\&\, e_1 \Rightarrow e_2$", which makes some sense for low-level hypotheses, cannot be exported to the domain of theories. As to the conditional "$e \Rightarrow h$", it makes no sense for scientific hypotheses, much less for scientific theories, since no set of data implies a hypothesis – if only because the latter may contain predicates that fail to occur in the former. Yet it is the declared aim of most systems of inductive logic to evaluate the degree of confirmation (or logical probability) of conditionals of this kind. Which explains why such theories are irrelevant to science. We may add that, so far, inductive logic has not faced the problem of devising reasonable measures of the degree of confirmation of quantitative theories: it has focused on stray hypotheses, and even here it has met with disaster. This does not prove, of course, that the very aim of inductive logic is chimerical, but just that the construction of systems of inductive logic relevant to science is a task before us.

A second consequence is that there can hardly be any *conclusive* evidence for or against a scientific theory. A set of data can, on occasion, confirm or refute a single hypothesis in an unambiguous way, but it is considerably less powerful in regard to a theory. While agreement between theory and experience – agreement over a given domain, that is – confirms the former, it does not point with certainty to the truth of the theory: it may indicate that both theory and data are sloppy, e.g. that compensating errors have crept into both. And disagreement between theory and experience cannot always be interpreted as a clean refutation of the former either. Confirmation and refutation, clear-cut though they can be in the case of single qualitative hypotheses – the case considered by both the inductivists and their critics – lose much of their edge in the case of quantitative theoretical predictions. This is not to say that scientific theories are impregnable to experience but rather that the process of their empirical test is complex and roundabout. (This renders the cogent and explicit, i.e. axiomatic formulation of theories the more valuable, for it facilitates the control of the assumptions.) The complex and often inconclusive character of empirical testing enhances the value of the nonempirical tests, which are ultimately tests for the global consistency of the whole body of scientific knowledge.

Inductivism and refutationism are then inadequate, for both restrict themselves to single hypotheses, both neglect the theoretical model that must be adjoined to a general theory in order to deduce testable consequences, and both accept the tenets that (*a*) only empirical tests matter and (*b*) the outcome of such tests is always clear-cut. However, the failure of the currently dominant philosophies of science should not thrust us into the arms of conventionalism or any other philosophical expression of cynicism. We are entitled to hope that some of our theories are internally and externally consistent and that they have at least a grain of truth even though we may not be able to prove either property beyond doubt. For this hope is not blind faith: it is founded on the performance of our theories – on their tested ability to get along with other theories, to solve old and new problems, to make novel predictions, and to make new experiences intelligible and even possible.

To summarise: theory and experience never meet head-on. They meet on an intermediate level once further theoretical and empirical elements, in particular theoretical models of both the thing concerned and the empirical arrangement, have been added. Even so, empirical tests are not always conclusive and they do not enable us to dispense with the nonempirical ones. To the extent to which all this is true, the dominant philosophies of science are inadequate. We must start afresh, keeping closer to actual scientific research than to the philosophical traditions.

NOTE

* Some paragraphs are reproduced from Bunge (1970d) with the permission of the editor and publisher.

BIBLIOGRAPHY

Abro, A. d': 1939, *The Decline of Mechanism*, Van Nostrand, New York.

Agassi, J.: 1964, in Bunge, M. (ed.), 1964a.

Ayer, A. J.: 1968, in *Démonstration, vérification, justification: Entretiens de l'Institut International de Philosophie*, Nauwelaerts, Louvain.

Bastin, E. W. (ed.): 1971, *Quantum Theory and Beyond*, Cambridge University Press.

Beck, G. and Nussenzweig, H. M.: 1958, *Nuovo Cimento* **9**, 1068.

Bergmann, P. G.: 1967, in Bunge (1967e).

Birkhoff, G. and v. Neumann, J.: 1936, *Annals of Math.* **37**, 823.

Bohm, D.: 1951, *Quantum Theory*, Prentice-Hall, Englewood Cliffs, N.J.

Bohm, D. and Bub, J.: 1966, *Rev. Mod. Phys.* **38**, 453.

Bohr, N.: 1934, *Atomic Theory and the Description of Nature*, Cambridge University Press.

Bohr, N.: 1958a *Atomic Physics and Human Knowledge*, Wiley, New York.

Bohr, N.: 1958b, in R. Klibansky (ed.), *Philosophy in the Mid-Century*, La Nuova Italia, Firenze.

Bourret, R.: 1964, *Phys. Letters* **12**, 323.

Boyer, T. H.: 1969a, *Phys. Rev.* **182**, 1374.

Boyer, T. H.: 1969b, *Phys. Rev.* **186**, 1304.

Bridgman, P. W.: 1927, *The Logic of Modern Physics*, Macmillan Co., New York.

Broglie, L. de: 1937, *Matière et lumière*, Gauthier-Villars, Paris.

Brush, S.: 1968, *Archive Hist. Exact Sci.* **5**, 5.

Bunge, M.: 1955a, *Brit. J. Phil. Sci.* **6**, 1, 141.

Bunge, M.: 1955b, *Methodos* **7**, 295.

Bunge, M.: 1956, *Am. J. Phys.* **24**, 272.

Bunge, M.: 1957, *Am. J. Phys.* **25**, 211.

Bunge, M.: 1959a, *Causality*, Harvard University Press, Cambridge, Mass.; Rev. ed.: 1963, The World Publ. Co., Cleveland and New York.

Bunge, M.: 1959b, *Metascientific Queries*, Charles C. Thomas, Publ., Springfield, Ill.

Bunge, M.: 1961a, *Am. J. Phys.* **29**, 518.

Bunge, M.: 1961b, *Phil. Sci.* **28**, 120.

Bunge, M.: 1962a, *Intuition and Science*, Prentice-Hall, Englewood Cliffs, N. J.

Bunge, M.: 1962b, *The Monist* **47**, 116.

Bunge, M.: 1963, *The Myth of Simplicity*, Prentice-Hall, Englewood Cliffs, N. J.

Bunge, M. (ed.): 1964a, *The Critical Approach*, Free Press, Glencoe, III.

Bunge, M.: 1964b, in Bunge (1964a).

Bunge, M.: 1965, *Dialectica* **19**, 195.

Bunge, M.: 1966, *Am. J. Phys.* **34**, 585.

Bunge, M.: 1967a, *Foundations of Physics*, Springer-Verlag, New York.

Bunge, M.: 1967b, *Rev. Mod. Phys.* **39**, 463.

Bunge, M.: 1967c, *Scientific Research*, 2 vols., Springer-Verlag, New York.

Bunge, M. (ed.): 1967d, *Delaware Seminar in the Foundations of Physics*, Springer-Verlag, New York.

Bunge, M. (ed.): 1967e, *Quantum Theory and Reality*, Springer-Verlag, New York.

Bunge, M.: 1967f, *Brit. J. Phil. Sci.* **18**, 265.
Bunge, M.: 1968a in I. Lakatos and A. Musgrave (eds.), *Problems in the Philosophy of Science*, North-Holland, Amsterdam.
Bunge, M.: 1968b, *Intern. J. Theor. Phys.* **1**, 205.
Bunge, M.: 1968c, *Revue Internationale de Philosophie* **23**, 16.
Bunge, M.: 1968d, *Philosophy of Science* **35**, 355.
Bunge, M.: 1969a, *Proc. XIVth Intern. Congress of Philosophy* **III**, Herder, Wien.
Bunge, M.: 1969b, *American Philosophical Quarterly Monograph*, No. 3, 62.
Bunge, M.: 1970a, *Can. J. Phys.* **48**, 1410.
Bunge, M.: 1970b, *Zeitschrift für allgemeine Wissenschaftstheorie* **1**, 196.
Bunge, M.: 1970c, in Weingartner and Zecha (1970).
Bunge, M.: 1970d, in Kiefer and Munitz (1970).
Bunge, M.: 1970e, *Studium Generale* **23**, 562.
Bunge, M.: 1971a, in Bastin (1971).
Bunge, M. (ed.): 1971b, *Problems in the Foundations of Physics*, Springer-Verlag, New York.
Bunge, M.: 1971c, in Bunge (1971b).
Bunge, M.: 1972a, *Method, Model, and Matter*, D. Reidel Publ. Co., Dordrecht.
Bunge, M. (ed.): 1972b, *Exact Philosophy: Problems, Methods, Goals*, D. Reidel Publ. Co., Dordrecht.
Bunge, M.: 1972c, in Bunge (1972b).
Carathéodory, C.: 1924, *Sitzber. Preuss. Acad. Wiss. Phys. – Math. Kl.*, 12.
Carnap, R.: 1950 *Logical Foundations of Probability*, University of Chicago Press.
Carnap, R.: 1958, *Introduction to Symbolic Logic and its Applications*, New York, Dover Publications, Inc.
Church, A.: 1962, in E. Nagel, P. Suppes and A. Tarski (eds.), *Logic, Methodology and Philosophy of Science*, Stanford University Press.
Coffa, J. A.: 1967, *J. Phil.* **64**, 500.
Daneri, A., Loinger, A., and Prosperi, G. M.: 1962, *Nuclear Physics* **33**, 297.
Destouches-Février, P.: 1951, *La structure des théories physiques*, Presses Universitaires de France, Paris.
Dirac, P. A. M.: 1958, *The Principles of Quantum Mechanics*, 4th ed., Clarendon Press, Oxford.
Dirac, P. A. M.: 1972, *Proc. Roy. Soc. Lond.* **A 328**, 1.
Duhem, P.: 1914, *La théorie physique*, 2nd ed., Rivière, Paris.
Edelen, D. G. B.: 1962, *The Structure of Field Space*, University of California Press, Berkeley and Los Angeles.
Everett III, H.: 1957, *Rev. Mod. Phys.* **29**, 454.
Feigl, H.: 1962, *Phil. Sci.* **29**, 39.
Feigl, H.: 1967, *The 'Mental' and the 'Physical'*, University of Minnesota Press, Minneapolis.
Feyerabend, P. K.: 1962, in H. Feigl and G. Maxwell (eds.), *Minnesota Studies in the Philosophy of Science*, III.
Feyerabend, P. K.: 1968, *Phil. Sci.* **35**, 309.
Fine, A.: 1968, *Phil. Sci.* **35**, 101.
Frank, P.: 1938, *Interpretations and Misinterpretations of Modern Physics*, Hermann & Cie, Paris.
Frank, P.: 1946, *Foundations of Physics*, University of Chicago Press.
Freudenthal, H.: 1970, *Synthese* **21**, 93.

Grad, H.: 1967, in Bunge 1967d.

Groenewold, H. J.: 1968, *Foundations of Quantum Theory*, preprint of the Institute for Theoretical Physics, Groningen University.

Hebb, D. O.: 1966, *A Textbook of Psychology*, 2nd ed., Philadelphia: W.B. Saunders.

Heitler, W.: 1963, *Man and Science*, Basic Books, New York.

Helmholtz, H. v.: 1879, *Die Tatsachen in der Wahrnehmung*, Verlag August Hirschwald, Berlin.

Hempel, C. G.: 1965, *Aspects of Scientific Explanation*, Free Press, New York.

Henkin, L., Suppes, P. and Tarski, A. (eds.): 1959, *The Axiomatic Method*, North-Holland, Amsterdam.

Hertz, H.: 1956, *The Principles of Mechanics*, transl. by P. E. Jones and J. T. Walley, Dover, New York.

Hesse, M.: 1966, *Models and Analogies in Science*, Notre Dame University Press, Notre Dame, Ind.

Hilbert, D.: 1909, *Math. Ann.* **67**, 355.

Hilbert, D.: 1912, *Phys. Zeits.* **13**, 1056.

Hilbert, D.: 1913, *Phys. Zeits.* **14**, 592.

Hilbert, D.: 1914, *Phys. Zeits.* **15**, 878.

Hilbert, D.: 1918, *Math. Ann.* **78**, 405.

Hilbert, D.: 1924, *Math. Ann.* **92**, 1.

Hooker, C. A.: 1972, in R. G. Colodny (ed.): 1972, *Paradigms and Paradoxes*, Pittsburgh University Press, Pittsburgh.

Houtappel, R. M. F., van Dam, H., and Wigner, E. P.: 1965, *Rev. Mod. Phys.* **37**, 595.

Jost, R.: 1965, *The General Theory of Quantized Fields*, Amer. Math. Soc., Providence, R. I.

Kiefer, H. and Munitz, M. (eds.): 1970, *Contemporary Philosophic Thought*, Vol. 2, State University of New York Press, Albany, N.Y.

Kuhn, T. S.: 1962, *The Structure of Scientific Revolutions*, University of Chicago Press.

Lakatos, I. (ed.): 1968, *The Problem of Induction*, North-Holland, Amsterdam.

Landau, L. and Lifshitz, E. M.: 1958, *Quantum Mechanics*, Pergamon Press, Oxford.

Landé, A.: 1965, *New Foundations of Quantum Mechanics*, Cambridge University Press.

Levi, B.: 1947, *Leyendo a Euclides*, Editorial Rosario, Rosario.

Ludwig, G.: 1967, In Bunge 1967e.

Margenau, H.: 1950 *The Nature of Physical Reality*, McGraw-Hill, New York.

Margenau, H. and Park, J.: 1967, in Bunge (1967e).

Mariwalla, K. H.: 1969, *Am. J. Phys.* **37**, 1281.

Marshall, T. W.: 1963, *Proc. Roy. Soc.* A **276**, 475.

Marshall, T. W.: 1965, *Proc. Cambridge Phil. Soc.* **61**, 537.

Maxwell, J. C.: 1871, *Proc. London Math. Soc.* **3**, 224.

McKinsey, J. C. C., Sugar, A. C., and Suppes, P.: 1953, *J. Rational Mech. Anal.* **2**, 253.

Meggers, W. F., Corliss, C. H., and Scribner, B. F.: 1961, *Tables of Spectral-Line Intensities*, National Bureau of Standards Monograph 32, Washington D. C.

Metzger, H.: 1926, *Les concepts scientifiques*, Alcan, Paris.

Miller, G. A.: 1967, *The Psychology of Communication*, Basic Books, New York.

Mises, R. von: 1951, *Positivism: A Study in Human Understanding* [transl. from the German ed., 1939], Harvard University Press, Cambridge, Mass.

Nagel, E.: 1961, *The Structure of Science*, Harcourt, Brace & World, New York.

Nagel, E.: 1970, in Kiefer and Munitz (1970).

Neumann, J. v.: 1932, *Mathematische Grundlagen der Quantenmechanik*, Springer-Verlag, Berlin. Engl. transl.: *Mathematical Foundations of Quantum Mechanics*, Princeton University Press, 1955.
Newton, R. G.: 1966, *Scattering Theory of Waves and Particles*, McGraw-Hill, New York.
Noll, W.: 1959, in Henkin *et al.*, 1959.
Pauli, W.: 1954, *Dialectica* **8**, 112.
Peña-Auerbach, L. de la: 1969, *J. Math. Phys.* **10**, 1620–1630.
Poincaré, H.: 1912, *Calcul des Probabilités*, 2nd ed., Gauthier-Villars, Paris.
Popper, K. R.: 1959, *Brit. J. Phil. Sci.* **10**, 25.
Popper, K. R.: 1967, In M. Bunge 1967e.
Popper, K. R.: 1968, *Nature* **219**, 682.
Reichenbach, H.: 1924, *Axiomatik der relativistischen Raum-Zeit Lehre*, Fr. Vieweg und Sohn, Braunschweig. Engl. transl.: 1969, *Axiomatization of the Theory of Relativity*, University of California Press, Berkeley and Los Angeles.
Reichenbach, H.: 1951, *The Rise of Scientific Philosophy*, University of California Press, Berkeley and Los Angeles.
Robinson, A.: 1956, *Complete Theories*, North-Holland, Amsterdam.
Rosenfeld, L.: 1953, *Science Progress*, No. 163, 393.
Rosenfeld, L.: 1961, *Nature* **190**, 384.
Rosenfeld, L.: 1964, in L. Infeld (ed.), *Proceedings on Theory of Gravitation*, Gauthier-Villars, Paris.
Russell, B.: 1940, *An Inquiry into Meaning and Truth*, George Allen & Unwin, London.
Salt, D.: 1971, *Foundations of Physics* **1**, 307.
Schiller, R.: 1967, in Bunge, M. (ed.), 1967d.
Schilpp, P. A. (ed.): 1971, *The Philosophy of Karl Popper*.
Schrödinger, E.: 1926, *Ann. Phys.* **79**, 489.
Schrödinger, E.: 1933, *Mémoires sur la mécanique ondulatoire*, Gauthier-Villars, Paris.
Seshu, S. and Reed, M. B.: 1961, *Linear Graphs and Electrical Networks*, Addison-Wesley Publishing Co., Reading, Mass.
Settle, T. W.: 1971, in Schilpp 1971.
Smart, J. J. C.: 1963, *Philosophy and Scientific Realism*, Routledge & Kegan Paul, London.
Smith, B. A.: 1967, *Science* **158**, 114.
Smoluchowski, M. von: 1918, *Naturwiss.* VI, 253.
Stapp, H. P.: 1971, *Phys. Rev.* **D 3**, 1303.
Strauss, M.: 1970. In Weingartner and Zecha 1970.
Suppes, P.: 1966, in *The Role of Axiomatics and Problem Solving in Mathematics*, Boston, Ginn and Co.
Suppes, P.: 1967, *Set-Theoretic Structures in Science*, Institute for Mathematical Studies in the Social Sciences, Stanford University.
Suppes, P.: 1969, *Studies in the Methodology and Foundations of Science*, D. Reidel Publ. Co., Dordrecht.
Tarski, A.: 1956, *Logic, Semantics, Metamathematics*, Clarendon Press, Oxford.
Tisza, L.: 1962, *Rev. Mod. Phys.* **35**, 151.
Truesdell, C.: 1966, *Six Lectures on Modern Natural Philosophy*, Springer-Verlag, New York.
Truesdell, C.: 1969, *Rational Thermodynamics*, McGraw-Hill, New York.
Truesdell, C. and Toupin, R.: 1960, in S. Flügge, (ed.) 'The Non-Linear Field Theories of Mechanics', *Handbook of Physics*, III/1, Springer-Verlag, Berlin–Göttingen–Heidelberg.

Vaihinger, H.: 1920, *Die Philosophie des Als Ob*, 4th ed., Meiner, Leipzig.
Weingartner, P. and Zecha, G. (eds.): 1970, *Induction, Physics and Ethics*, D. Reidel, Dordrecht.
Wheeler, J. A.: 1957, *Rev. Mod. Phys.* **29**, 463.
Wightman, A. S.: 1956, *Phys. Rev.* **101**, 860.
Wigner, E.: 1962, in I. J. Good (ed.) *The Scientist Speculates*, Heinemann, London.
Woodger, J. H.: 1937, *The Axiomatic Method in Biology*, Cambridge University Press.
Woodger, J. H.: 1952, *Biology and Language*, Cambridge University Press.
Wright, G. H. von: 1957, *The Logical Problem of Induction*, 2nd. ed., Basil Blackwell, Oxford.

INDEX OF NAMES

INDEX OF SUBJECTS

SYNTHESE LIBRARY

Monographs on Epistemology, Logic, Methodology,
Philosophy of Science, Sociology of Science and of Knowledge, and on the
Mathematical Methods of Social and Behavioral Sciences

Editors:

DONALD DAVIDSON (Rockefeller University and Princeton University)
JAAKKO HINTIKKA (Academy of Finland and Stanford University)
GABRIËL NUCHELMANS (University of Leyden)
WESLEY C. SALMON (Indiana University)

‡SÖREN STENLUND, *Combinators, λ-Terms and Proof Theory.* 1972, 184 pp. Dfl. 40,—
‡DONALD DAVIDSON and GILBERT HARMAN (eds.), *Semantics of Natural Language.*
1972, X + 769 pp. (Cloth) Dfl. 95,—
(Paper) Dfl. 45,—
‡STEPHEN TOULMIN and HARRY WOOLF (eds.), *Norwood Russell Hanson: What I Do
Not Believe, and Other Essays.* 1971, XII + 390 pp. Dfl. 90,—
‡ROGER C. BUCK and ROBERT S. COHEN (eds.), *PSA 1970. In Memory of Rudolf Carnap.*
Boston Studies in the Philosophy of Science. Volume VIII (ed. by Robert S. Cohen
and Marx W. Wartofsky). 1971, LXVI + 615 pp. Dfl. 120,—
‡YEHOSHUA BAR-HILLEL (ed.), *Pragmatics of Natural Languages.* 1971, VII + 231 pp.
Dfl. 50,—
‡ROBERT S. COHEN and MARX W. WARTOFSKY (eds.), *Boston Studies in the Philosophy
of Science.* Volume VII: *Milič Čapek: Bergson and Modern Physics.* 1971, XV + 414 pp.
Dfl. 70,—
‡CARL R. KORDIG, *The Justification of Scientific Change.* 1971, XIV + 119 pp. Dfl. 33,—
‡JOSEPH D. SNEED, *The Logical Structure of Mathematical Physics.* 1971, XV + 311 pp.
Dfl. 70,—
‡JEAN-LOUIS KRIVINE, *Introduction to Axiomatic Set Theory.* 1971, VII + 98 pp.
Dfl. 28,—
‡RISTO HILPINEN (ed.), *Deontic Logic: Introductory and Systematic Readings.* 1971,
VII + 182 pp. Dfl. 45,—
‡EVERT W. BETH, *Aspects of Modern Logic.* 1970, XI + 176 pp. Dfl. 42,—
‡PAUL WEINGARTNER and GERHARD ZECHA (eds.), *Induction, Physics, and Ethics,
Proceedings and Discussions of the 1968 Salzburg Colloquium in the Philosophy of
Science.* 1970, X + 382 pp. Dfl. 65,—
‡ROLF A. EBERLE, *Nominalistic Systems.* 1970, IX + 217 pp. Dfl. 42,—
‡JAAKKO HINTIKKA and PATRICK SUPPES, *Information and Inference.* 1970, X + 336 pp.
Dfl. 60,—
‡KAREL LAMBERT, *Philosophical Problems in Logic. Some Recent Developments.* 1970,
VII + 176 pp. Dfl. 38,—
‡P. V. TAVANEC (ed.), *Problems of the Logic of Scientific Knowledge.* 1969, XII + 429 pp.
Dfl. 95,—
‡ROBERT S. COHEN and RAYMOND J. SEEGER (eds.), *Boston Studies in the Philosophy of
Science.* Volume VI: *Ernst Mach: Physicist and Philosopher.* 1970, VIII + 295 pp.
Dfl. 38,—
‡MARSHALL SWAIN (ed.), *Induction, Acceptance, and Rational Belief.* 1970, VII + 232 pp.
Dfl. 40,—

‡Nicholas Rescher *et al.* (eds.), *Essays in Honor of Carl G. Hempel. A Tribute on the Occasion of his Sixty-Fifth Birthday.* 1969, VII + 272 pp. Dfl. 50,—

‡Patrick Suppes, *Studies in the Methodology and Foundations of Science. Selected Papers from 1911 to 1969.* 1969, XII + 473 pp. Dfl. 72,—

‡Jaakko Hintikka, *Models for Modalities. Selected Essays.* 1969, IX + 220 pp. Dfl. 34,—

‡D. Davidson and J. Hintikka (eds.), *Words and Objections: Essays on the Work of W. V. Quine.* 1969, VIII + 366 pp. Dfl. 48,—

‡J. W. Davis, D. J. Hockney and W. K. Wilson (eds.), *Philosophical Logic.* 1969, VIII + 277 pp. Dfl. 45,—

‡Robert S. Cohen and Marx W. Wartofsky (eds.), *Boston Studies in the Philosophy of Science,* Volume V: *Proceedings of the Boston Colloquium for the Philosophy of Science 1966/1968.* 1969, VIII + 482 pp. Dfl. 60,—

‡Robert S. Cohen and Marx W. Wartofsky (eds.), *Boston Studies in the Philosophy of Science,* Volume IV: *Proceedings of the Boston Colloquium for the Philosophy of Science 1966/1968.* 1969, VIII + 357 pp. Dfl. 72,—

‡Nicholas Rescher, *Topics in Philosophical Logic.* 1968, XIV + 347 pp. Dfl. 70,—

‡Günther Patzig, *Aristotle's Theory of the Syllogism. A Logical-Philological Study of Book A of the Prior Analytics.* 1968, XVII + 215 pp. Dfl. 48,—

‡C. D. Broad, *Induction, Probability, and Causation. Selected Papers.* 1968, XI + 296 pp. Dfl. 54,—

‡Robert S. Cohen and Marx W. Wartofsky (eds.), *Boston Studies in the Philosophy of Science.* Volume III: *Proceedings of the Boston Colloquium for the Philosophy of Science 1964/1966.* 1967, XLIX + 489 pp. Dfl. 70,—

‡Guido Küng, *Ontology and the Logistic Analysis of Language. An Enquiry into the Contemporary Views on Universals.* 1967, XI + 210 pp. Dfl. 41,—

*Evert W. Beth and Jean Piaget, *Mathematical Epistemology and Psychology.* 1966, XXII + 326 pp. Dfl. 63,—

*Evert W. Beth, *Mathematical Thought. An Introduction to the Philosophy of Mathematics.* 1965, XII + 208 pp. Dfl 37,—

‡Paul Lorenzen, *Formal Logic.* 1965, VIII + 123 pp. Dfl. 26,—

‡Georges Gurvitch, *The Spectrum of Social Time.* 1964, XXVI + 152 pp. Dfl. 25,—

‡A. A. Zinov'ev, *Philosophical Problems of Many-Valued Logic.* 1963, XIV + 155 pp. Dfl. 32,—

‡Marx W. Wartofsky (ed.), *Boston Studies in the Philosophy of Science.* Volume I: *Proceedings of the Boston Colloquium for the Philosophy of Science, 1961–1962.* 1963, VIII + 212 pp. Dfl. 26,50

‡B. H. Kazemier and D. Vuysje (eds.), *Logic and Language. Studies dedicated to Professor Rudolf Carnap on the Occasion of his Seventieth Birthday.* 1962, VI + 256 pp. Dfl. 35,—

*Evert W. Beth, *Formal Methods. An Introduction to Symbolic Logic and to the Study of Effective Operations in Arithmetic and Logic.* 1962, XIV + 170 pp. Dfl. 35,—

*Hans Freudenthal (ed.), *The Concept and the Role of the Model in Mathematics and Natural and Social Sciences. Proceedings of a Colloquium held at Utrecht, The Netherlands, January 1960.* 1961, VI + 194 pp. Dfl. 34,—

‡P. L. Guiraud, *Problémes et méthodes de la statistique linguistique.* 1960, VI + 146 pp. Dfl. 28,—

*J. M. Bocheński, *A Precis of Mathematical Logic.* 1959, X + 100 pp. Dfl. 23,—

SYNTHESE HISTORICAL LIBRARY

Texts and Studies
in the History of Logic and Philosophy

Editors:

N. KRETZMANN (Cornell University)
G. NUCHELMANS (University of Leyden)
L. M. DE RIJK (University of Leyden)

‡LEWIS WHITE BECK (ed.), *Proceedings of the Third International Kant Congress.* 1972,
XI + 718 pp.   Dfl. 160, —
‡KARL WOLF and PAUL WEINGARTNER (eds.), *Ernst Mally: Logische Schriften.* 1971,
X + 340 pp.   Dfl. 80, —
‡LEROY E. LOEMKER (ed.), *Gottfried Wilhelm Leibnitz: Philosophical Papers and Letters.*
A Selection Translated and Edited, with an Introduction. 1969, XII + 736 pp.
                                                                                  Dfl. 125, —
‡M. T. BEONIO-BROCCHIERI FUMAGALLI, *The Logic of Abelard.* Translated from the
Italian. 1969, IX + 101 pp.   Dfl. 27, —

Sole Distributors in the U.S.A. and Canada:

*GORDON & BREACH, INC., 440 Park Avenue South, New York, N.Y. 10016
‡HUMANITIES PRESS, INC., 303 Park Avenue South, New York, N.Y. 10010